The Scientific Story of Creation

The Scientific Story of Creation

Adrian Bjornson

Addison Press
Woburn, Massachusetts
www.olduniverse.com

Published by: **Addison Press**
  400 West Cummings Park
  PMB 1725-111
  Woburn, MA 01801
  www.olduniverse.com

Copyright © 2002 by Addison Press

All rights reserved under International and Pan-American Copyright Conventions. No part of this book may be reproduced by any mechanical, photographic, or electronic process, nor may it be stored in a retrieval system, transmitted, or otherwise copied for public or private use, without written permission of Addison Press.

Printed in the United States of America

Publisher's Cataloguing-in-Publication data
  (Provided by Quality books, Inc.)

Bjornson, Adrian
  The scientific story of creation / Adrian Bjornson.
  -- 1st ed.
  p. cm.
  Includes bibliographical references and index.
  ISBN 09703231-2-3
  1. Cosmology--Popular works. 2. Creation
I. Title.

QB982.B567 2002   523.1
                     QB102-200196

This book is dedicated to

Professor Huseyin Yilmaz

*Who has devoted his life to develop his theory of gravity,
which extends the Einstein relativity concepts
to achieve the goal that Albert Einstein sought.*

Acknowledgements

I am grateful for the patient assistance that Prof. Huseyin Yilmaz has provided in explaining his *Theory of Gravity* and the principles of the Einstein *General Theory of Relativity*.

I thank William C. Keel of the University of Alabama for his excellent photograph of the M51 Whirlpool Galaxy with its companion galaxy NGC 5195. This was taken on the 1.1-meter Hall telescope at the Lowell Observatory. I have used this photograph on the front cover and in Figure 1-1.

Foreword

Issues to Be Presented

This book examines the scientific evidence concerning Creation. Scientists have derived a clear picture of the creation of life on earth, since the earth was formed 4.6 billion years ago. They have also shown how our earth and our sun were created, how other stars were created, and how the stars, including our sun, will eventually die.

But when we consider the creation of our whole universe, the picture becomes confusing and controversial. It was much simpler in the 1950's. Then there were two competing concepts: the Big Bang theory of nuclear physicist George Gamow, and the Steady-State Universe theory of astrophysicist Fred Hoyle. The Steady-State Universe theory assumes that the universe age is infinite and that diffuse matter is continuously created to compensate for the Hubble expansion of the universe.

Gamow assumed that the universe began with an enormous Big Bang explosion billions of years ago. He postulated that at the time of creation the universe had the density of a neutron star, which is the maximum density allowable under normal physical laws. With the Gamow postulate, the observable universe at the instant of the Big Bang would have just fit inside the orbit of the planet Mars.

Modern Big Bang cosmologists have discarded the Gamow concept. They postulate that the universe began as a *"singularity"*, which ideally means an infinite density of matter. Astrophysicist P. James E. Peebles, acclaimed as the *"father of modern cosmology"*, predicted that the observable universe was initially *"smaller than a dime"*. This *singularity* concept was derived from the Einstein General theory of Relativity.

Yet Einstein absolutely rejected a *singularity* prediction associated with the black hole, which was derived from his theory in a 1939 paper, and in 1945 he rejected the concept that the universe was created as a *singularity*. **Einstein insisted that, "singularities do not exist in physical reality". After that 1939 paper, no scientist claimed that the Einstein theory predicted a singularity while Einstein was alive.**

To make sense from these issues, the reader must understand the

Einstein General theory of Relativity. Although the equations of the Einstein theory are very complicated, this book shows that the physical principles behind the mathematics can be explained in a simple yet scientifically accurate manner.

In 1958, Huseyin Yilmaz published in the prestigious *Physical Review* a new theory that is a refinement of the Einstein theory. It incorporates the principles of the Einstein theory but eliminates its *singularity* predictions. When applied to cosmology, the Yilmaz theory predicts a universe that is similar to the Steady-State Universe theory but does not have its limitations.

Simplicity of Presentation

This book addresses deep theoretical issues in a simple physical manner that is accurate and detailed. To achieve this end, the book has five appendices containing supplementary material, which most readers can ignore. In addition, the book references other documents by the author that contain more complicated supporting information.

The supporting documents by the author are listed as Refs. [1, 2, 3] of the Bibliography. The titles of these are

Believe [1]: *A Universe that We Can Believe*
Created [2]: *How Was Our Universe Created?*
Addendum [3]: *Addendum* section of website, www.olduniverse.com

In this book, these documents are referenced as *Believe* [1], *Created* [2], and *Addendum* [3].

The document *Addendum* [3] is available at no cost on the website. This material is directed toward those with scientific training and assumes the knowledge of calculus.

The documents *Believe* [1] and *Created* [2] are written for the general reader, but are more detailed than this book. These documents are described in the world-wide website www.olduniverse.com, which shows how they can be purchased. The document *Believe* [1] has a number of appendices containing detailed analyses.

With the appendices and these documents to provide supporting information, the book is very readable even though it addresses profound scientific issues

At the end of the book is a Glossary, which defines terms used in this book. The Glossary should eliminate confusion when the book uses words that are unfamiliar to the reader.

Contents

Dedication and acknowledgements	v
Foreword	vii
Preface	xiv
1. Introduction	**1**
The eternal questions of creation	1
Searching for answers	2
The evolution of life on earth	2
Our sun and our Milky Way galaxy	4
The mystery of cosmology	4
The Hubble expansion of the universe	*4*
Serious questions concerning the Big Bang theory	*5*
Alternate explanations for the Hubble expansion	6
The Hubble expansion is apparent	*6*
The Steady State Universe theory	*7*
The singularity concept	9
The Einstein General theory of Relativity	9
The Yilmaz refinement of the Einstein theory	10
Einstein's rejection of the singularity concept	11
The Yilmaz cosmology model	12
The genius of Albert Einstein	13
Our scientific story of creation	13
2. The creation of life on earth	**14**
Early life	14
The first animals	15
Development of fishes	17
Amphibians invade the land	18
Spread of plants over the land	19
The reign of the reptiles	21
The Synapsids	*22*
The Dinosaurs	*23*
The slow rise of the mammals	24
The ascent of humans	25
Australopithecus and the Homo genus	*25*
Hunting capabilities of early man	*26*
From Homo Erectus to anatomically modern humans	*27*
The development of sophisticated language	*28*
How the Homo genus evolved	*30*
Was there a divine spark in the development of humanity?	*32*
The growth of civilization	*32*
Beyond the earth	33

3. The creation of our stars — 34
- Our Milky Way galaxy — 34
  - *Enormous distances to the stars* — 36
  - *The search for intelligent extraterrestrial life* — 39
- The birth of our sun and our earth — 40
- The life and death of our sun and similar stars — 43
  - *Contraction to a white dwarf* — 44
  - *Life cycles of other stars* — 45
  - *The meaning of absolute magnitude* — 46
  - *Our search for stars with intelligent life* — 47
- The supernova — 50
  - *The neutron star* — 50
  - *The pulsar* — 52
- The black hole — 52
- Radiation from an ideal blackbody — 56

4. The creation of our universe — 59
- *The view of our universe in 1900* — 59
- *The local group* — 60
- The Hubble expansion of our universe — 60
  - *Concept of the Hubble expansion* — 60
  - *The apparent age of the universe* — 62
  - *Measurement of galaxy distance* — 63
  - *Measurement of parallax* — 63
  - *The Cepheid variable stars* — 64
  - *How Edwin Hubble measured the galaxy distances* — 64
  - *Modern measurements of the Hubble constant* — 65
- The Big Bang concept — 66
  - *Material contained in our universe* — 66
  - *The neutron star model of universe creation* — 68
  - *George Gamow, the father of the Big Bang* — 70
  - *The singularity model of universe creation* — 72
- Cosmic microwave background radiation — 76
- The Quasar — 79
  - *The discovery of the quasar* — 79
  - *Quasar observations of Halton Arp* — 80
- Implications of gravitational theory — 82

5. Newton's theory of gravity — 83
- Development of Newton's theory — 83
- The Heliocentric theory of Copernicus — 84
  - *Galileo and his telescope* — 85
  - *The Kepler laws of planetary orbits* — 86
- Galileo's measurements of falling bodies — 87
- Calculation of the acceleration of gravity — 89
- The orbits of planets around the sun — 90
- Orbit of the moon around the earth — 93
- Engineering use of Newton's laws — 93
- Why are astronauts weightless? — 94
- How Cavendish weighed the earth — 95
- Coordinates to specify a vector — 96

6. The nature of light — 100
- What is a light wave? — 100
  - Mechanical waves — 100
  - Electromagnetic waves — 101
  - Meaning of electric and magnetic fields — 102
  - Principle of an electromagnetic wave — 103
- Early concepts of light — 104
- Galileo and Kepler — 104
- The optical discoveries of Isaac Newton — 105
- The wave property of light — 108
- The electromagnetic wave concept — 112
- Search for the velocity of the luminiferous aether — 113
  - The Michelson-Morley Experiment — 115
  - The contraction hypothesis — 116
  - The Lorentz transformation — 116
  - The Einstein Relativity principle — 117
  - Reaction to the Einstein Relativity principle — 118

7. Einstein Special theory of Relativity — 119
- Measuring the speed of sound — 119
- Measuring the speed of light — 120
- The Einstein theory of Relativity — 122
  - The principles of Relativity — 123
  - Explanation of constancy of the speed of light — 124
  - Replacing the observer by a set of coordinates — 125
  - Four dimensionality of space and time — 126
- Variation of mass of an object — 126
- Converting matter into energy — 128
- The principle of covariance — 129

8. The Einstein theory of gravity — 131
- Generalizing the Relativity principle — 131
- Applying the equivalence principle — 134
  - Equivalence between acceleration and gravity — 134
  - Redshift produced by gravity — 135
  - Effect of gravity on clock rate — 136
  - Other effects due to gravity and acceleration — 137
- Application of tensor analysis to General Relativity — 137
  - The metric tensor — 138
  - Converting form of a tensor — 140
  - The Ricci and Einstein curvature tensors — 140
  - The energy-momentum tensor — 141
  - The Einstein gravitational field equation — 141
  - Outline of Einstein theory calculations — 141
  - The Schwartzschild solution — 143
  - Computer solutions of the Einstein theory — 145

9. The Yilmaz theory of gravity — 146
- Derivation of the Yilmaz solution — 146
  - The elements of the metric tensor — 146
  - The Yilmaz gravitational field equation — 147
- The general time-varying Yilmaz theory — 148

Discussion of the Yilmaz theory	148
Reason for opposition to the Yilmaz theory	150
Consistency with quantum mechanics	150

10. Applying the Einstein and Yilmaz theories — 151

Relativistic effects produced by gravity	151
The normalized relativistic mass m	*152*
Effect of gravity on the speed of light	153
The black hole	154
Second limit to the Schwartzschild solution	*156*
Gravitational effect on distance and clock rate	157
Effect of gravitational field on wavelength	*158*
The quasar redshift	159

11. The quasar — 160

Summary of quasar characteristics	160
Quasar observations of Halton Arp	161
Statistical evidence given by Halton Arp	*162*
Principle for calculating probabilities	*164*
How much power does a quasar radiate?	*165*
Diameters of the associated galaxies	*165*
Galaxies with intrinsic redshift	*166*
Possible explanations for intrinsic redshift	166
Intrinsic redshift of quasar 3C48	*166*
The implications of forbidden spectral lines	*167*
Explanations for intrinsic quasar redshifts	*168*
Confusion in quasar research	*168*

12. Evidence against the Big Bang — 169

The editorial of Geoffrey Burbidge	169
Eric Lerner and Nobel Laureate Hannes Alfven	170
The Big Bang age dilemma	*171*
Mythological philosophy of Big Bang research	*172*
Effect of the computer on cosmological studies	175
Quasar studies by astronomer Halton Arp	177
Lack of scientific objectivity in astronomy today	178
Cosmic microwave radiation	179

13. Weaknesses of the Einstein theory — 180

Does not achieve a two-body solution	180
Professor Carroll O. Alley	*180*
The single-body Schwartzschild solution	*181*
The analysis of Professor Alley	*182*
Reason for failure to achieve a two-body solution	182
The Einstein, Ricci, and Energy-Momentum Tensors	*183*
The Riemann tensor	*183*
Conservation of matter-plus-energy	184
Multiple solutions for the Einstein theory	185
The Einstein theory is not rigorous	185
Variation of speed of light with direction	186

14. The Yilmaz cosmology model — 187

Description of Yilmaz cosmology model	188
Reduction of speed of light, clock rate, and spatial dimensions	*188*

The Hubble expansion of the universe	*191*
How can gravity make the universe expand?	*194*
Creation of matter	*195*
Cosmic microwave background radiation	196
Uniqueness of cosmology model predictions	196

15. A believable picture of our universe — 197

Viable models of our universe	197
The implications of the Yilmaz cosmology model	198
Matter derived from gravitational waves	*199*
The local expansion of the universe	*200*
The second law of thermodynamics	*200*
The contents of the universe	203
The "observable" Yilmaz universe	*203*
Is the universe infinite?	*204*
Religious and philosophical implications of our picture of the universe	205
The Biblical story of Creation	*205*
Our picture of the universe	*206*

16. The Genius of Albert Einstein — 207

Einstein's discovery of Relativity	207
The basic Relativity principle	*207*
Generalizing the Relativity principle	*208*
The Yilmaz refinement of the Einstein theory	209
Other achievements of Albert Einstein	210
Einstein's search for a unified field theory	211

Appendices — 212

A. The Marmet redshift effect — 213

B. Analysis of quasar 3C48 — 215

Analysis by Greenstein and Schmidt	215
Basic analysis	*215*
Rapid variation of quasar brightness	*218*
Other data on quasar 3C48	218

C. Details of Yilmaz cosmology model — 220

Cosmic microwave background radiation	220
Density of matter in the universe	224

D. The meaning of a tensor — 226

Tensor to specify stress within a body	226
Tensors in relativity theory	229

E. Relativity analyses — 230

Formulas for Schwartzschild and Yilmaz solutions	230
The meaning of relativistic gravitational potential	231
Derivation of Yilmaz theory	233
Relativistic Doppler shift	236
Einstein pseudo-tensor for the gravitational field	237
Conservation of matter-plus-energy	238
Einstein's rejection of the "Big Bang" singularity	239

Glossary — 240
Bibliography — 244
Index — 248-254

Preface

We begin our story of Creation by studying the evolution of life on earth. Our earth was created 4.6 billion years ago. One billion years later the first definite signs of life appeared in the form of cyanobacteria, which are microscopic cells that implement photosynthesis. It took about 2 billion more years before seaweed appeared, and another billion years before animal life emerged in our oceans, 600 million years ago.

We will trace the evolution of vertebrates, from simple chordates to modern mammals, which culminated in the creation of humans. Although anatomically modern humans have existed for 100 thousand years, modern human behavior began suddenly, 40 thousand years ago, when humans started acting in a radically new manner.

Next we raise our eyes to the heavens to see how the stars were created, including our sun. Gas and dust coalesced to form our sun, in which nuclear fusion was ignited 5 billion years ago. "Let there be Light, and there was Light." A disk of gas and dust surrounded the sun and produced our solar system. This disk seems to have been essential in the formation of our sun, and so the creation of a solar system around a star is probably a normal development. Hence there should be many stars with planets that are similar to our earth. We will examine the possibility that nearby stars have planets that contain advanced life.

We will see how our sun and other stars will grow old and eventually die. Most stars end their lives quietly, but very large stars die in an enormous explosion called a *supernova*. The violent supernova leaves behind an extremely dense body called a *neutron star*.

We will examine our Milky Way galaxy, with its 100 billion stars, and then we will look to our enormous universe that lies beyond, with its billions of galaxies. In 1929, astronomer Edwin Hubble discovered that our universe seems to be flying apart. It appears to have emerged from a single point in a tremendous explosion billions of years ago. This leads to the ultimate mystery, "How was our universe created?"

Preface **xv**

The Gamow Big Bang Concept

George Gamow was a noted physicist who worked on the Manhattan atomic bomb project. In the decade after World War II, Gamow publicized the concept that our universe began as a highly dense mass that exploded with a Big Bang. According to recent astronomical data, this explosion would have occurred about 15 billion years ago. Gamow proposed that our universe was initially compressed to the density of a *neutron star*, which he considered to have the greatest density of matter that is physically possible. A *neutron star* consists entirely of tightly packed neutrons and weighs 300 million tons per cubic centimeter.

As defined by the Big Bang theory, the observable universe has a radius of 15 billion light years. If this observable universe were compressed to the density of water, it would be 3 light years in diameter. If it were then compressed further to the density of a neutron star, it would just fit within the orbit of the planet Mars. According to the Gamow Big Bang theory, this was the size of the observable universe at the instant of the Big Bang.

The Big Bang Singularity

Since the time that computers became widely available in the mid 1960's, hundreds of scientists have devoted their careers to computer studies using the extremely complicated equations of the Einstein General theory of Relativity. Because of the great awe of Einstein, this research has been very rewarding professionally. Cosmology is the only area to which these computer studies can be applied, and so this has led to an enormous research effort on the Big Bang theory.

Computer studies of the Einstein theory have predicted that our universe began as a "singularity" at the instant of the Big Bang. Ideally a singularity is a body squeezed to zero size with infinite mass density. However, it is usually interpreted to mean an extremely compact body that has an enormous density of matter.

The January 2001 issue of *Scientific American* (p. 37) acclaimed astrophysicist P. James E. Peebles to be the *"father of modern cosmology"*. In the October 1994 *Scientific American* (p. 53), an article by Peebles and others described the Big Bang theory and claimed that the observable universe was initially *"concentrated in a region smaller than a dime"* at the instant of the Big Bang.

The Big Bang theory underwent a radical change when it evolved from the Gamow Big Bang theory to the modern Big Bang theory with its *singularity* postulate. ***In this change the initial universe was squeezed from the size of the Mars orbit to the size of a dime.*** What scientific evidence is given to support this magical compression of matter? The prediction is based solely on computer studies of the Einstein General theory of Relativity.

The concept that the Einstein theory predicts a singularity was first proposed in a 1939 paper, which claimed that a star with sufficient density to form what was later called a *black hole* would contract "indefinitely". The star would shrink to become a "singularity" having an infinite density of matter. ***Einstein absolutely rejected the black-hole singularity; he insisted that, "singularities do not exist in physical reality". After that rebuttal by Einstein, no scientist claimed that General Relativity predicted a singularity while Einstein was alive.***

Skepticism Concerning the Big Bang

For many years most astronomers have been informing the public with complete confidence that our universe was created with a Big Bang about 15 billion years ago. However, cosmologists are continually inventing artificial postulates to explain the many conflicts of the Big Bang theory with physical evidence. The following comment in the August 2001 *Scientific American* (p. 14) shows that there is growing skepticism about the validity of Big Bang research:

> *"Whenever Scientific American runs an article on cosmology, we get letters complaining that cosmology isn't a science, just unconstrained speculation".*

Understanding the Einstein Theory

To investigate cosmology, we must understand the Einstein General theory of Relativity. This book describes the Einstein theory in a simple physical manner that can be readily comprehended by the average reader.

The principles that Einstein established in developing his General theory of Relativity were very sound. However, Einstein derived the tensor formula that specifies his theory in an intuitive manner. That tensor formula has a flaw, and the physically impossible singularity predictions claimed by modern Big Bang cosmologists are due to that flaw.

Huseyin Yilmaz studied the Einstein theory as part of his PhD research at the Massachusetts Institute of Technology. He examined an approximate calculation that Einstein had made and realized that he could implement it exactly. This yielded an exact solution of Einstein's Relativity principles and resulted in the Yilmaz theory of gravity. Since the Yilmaz theory applies the principles of the Einstein theory, it is not a new theory; it a refinement of the Einstein theory.

Yilmaz published the first paper on his gravitational theory in 1958 in the prestigious *Physical Review*, and since then has published numerous scientific papers to extend his theory. The Yilmaz theory has eliminated the physically impossible singularity predictions that have been derived from the Einstein theory. Since the Yilmaz theory is a refinement of the Einstein theory, it has proven that the basic principles of the Einstein theory are inconsistent with singularities.

Big Bang cosmologists reject the Yilmaz theory, which refutes their singularity predictions. The Yilmaz theory is easy to apply, and so would reduce to obsolescence the elaborate computer techniques they have developed to solve the Einstein equations.

The Yilmaz Cosmology Model

In 1948, the famous astrophysicist Fred Hoyle proposed the *Steady State Universe* theory, which postulates that our universe is infinitely old and that diffuse matter is continually created throughout the universe to compensate for the universe expansion. This cosmology theory received strong support until *cosmic microwave radiation* was discovered in 1965, which had been predicted by Gamow. This radiation was claimed to be the cooled relic of optical radiation emitted 300,000 years after the Big Bang. The Steady-State Universe theory could not explain this cosmic microwave radiation, and so it fell into disfavor.

The Yilmaz theory yields a cosmological model that is similar to the Steady State Universe theory but does not have its limitations. Appendix C shows that the Yilmaz cosmology model predicts cosmic microwave radiation much more accurately than the Big Bang theory.

Since modern Big Bang theories are based on physically impossible singularity postulates, they are not realistic explanations of cosmology. This leaves us with the Gamow Big Bang theory and the Yilmaz cosmology model as viable hypotheses for how our universe was created.

Figure 1-1: The M51 Whirlpool galaxy, which is 35 million light years away, resembles our own Milky Way galaxy. Its smaller companion galaxy is NGC5195.

Chapter 1

Introduction

The Eternal Questions of Creation

Since the dawn of historical time humans have asked, "How were we created?" and, "How was our world created?" The Bible starts with its answer: "In the Beginning, God created the heavens and the earth". Now we ask the scientists for their answers.

To gain insight into this mystery, look up at the night sky. There is probably nothing more beautiful than the stars on a clear moonless night in a rural location that is far from city lights. The most inspiring feature is that pale white pathway across the sky called the Milky Way. Many people today have never seen the Milky Way, because it is obscured by sky glow due to the reflection of electric lights from the atmosphere. The Milky Way becomes particularly exciting when we realize what it represents. We are looking at billions and billions of distant stars within our Milky Way galaxy.

Our Milky-Way galaxy is similar to the M51 Whirlpool galaxy in Fig. 1-1 and on the front cover. It has a disk-like shape, with arms spiraling out from a central nucleus. Our sun is a star located 2/3 of the distance from the center to the circumference, and lies within a spiral arm. Our Milky Way galaxy, which contains 100 billion stars, is so vast that light takes one hundred thousand years to travel across it.

Prior to 1900 it was generally believed that our galaxy was the whole universe. Then astronomers developed means of measuring stellar distances and discovered that many of the fuzzy astronomical objects called "nebulae" were actually distant galaxies containing billions of stars like our own Milky Way. Our Milky Way galaxy is merely one out of billions of galaxies that comprise our universe. The size of our universe seems beyond comprehension.

Yet there is something even more unbelievable about our universe. In 1929, astronomer Edwin Hubble discovered that our universe is expanding. Except for a few close galaxies, he found that all galaxies are moving away from us at velocities approximately proportional to distance. Our whole universe seems to be flying apart. What can this mean? What does it tell us about the creation of our universe?

Searching for Answers

Most astronomers today are convinced that our whole universe was created as an extremely dense mass that exploded with a Big Bang about 15 billion years ago. For many years, this Big Bang theory has been proclaimed to be fact in scientific publications. Nevertheless, many responsible voices are questioning this theory, as demonstrated by the following comment in the August, 2001 *Scientific American* (p. 14):

> "Whenever Scientific American runs an article on cosmology, we get letters complaining that cosmology isn't a science, just unconstrained speculation."

This book examines the scientific evidence concerning creation. Scientists have developed a clear picture of the creation of life on earth, since the earth was formed 4.6 billion years ago. They have also shown how our earth and our sun were created, how other stars were created, and how the stars, including our sun, will eventually die.

But when we consider the creation of our whole universe, the picture becomes murky. Like the Bible, the answers of cosmological science appear to be based more on faith than on scientific evidence.

In our scientific study of Creation, we will separate the information that has strong scientific support from that which is questionable. We start by examining the development of life on earth, for which the basic story seems to be clear, although many questions of detail remain.

The Evolution of Life on Earth

Science has learned a great deal of how life has developed on earth, and of how humanity has evolved. There is strong evidence that our earth was formed as a molten body 4.6 billion years ago. The first clear sign of life occurred one billion years later. Chapter 2 traces the development of life on earth from microscopic organisms down to the complex life that we observe today. This scientific story of the evolution of life culminated in the emergence of humanity.

Since Darwin's theory of evolution was presented in 1857, many have disputed his theory. After all, we cannot observe the detailed steps of evolution, because the fossil evidence is sketchy, and evolution involved a countless series of small changes.

However, DNA research has now proven the strong biological similarity of different species. When we compare the very similar DNA codes in the cells of humans and animals, it seems impossible to take the position that the cells in our human bodies are not biologically related to the cells in these different species.

Does this mean that there was no divine force involved in the development of humanity? The basic Darwin theory postulates that the evolutionary process was the result of random mutations, with natural selection choosing those changes that were beneficial? But is this all that there was? To approach this issue in a more positive manner, we ask the reverse question. Is there any evidence in the evolution of humanity that there was something more than random biological evolution?

As Chapter 2 will show, a very strange and profound event occurred in human development about 40 thousand years ago. Fossils of anatomically modern humans appeared 100 thousand years ago. For at least 60 thousand years, Neanderthal "cave men" and anatomically modern humans made the same stone tools, and lived essentially the same lives, as far as the fossil evidence can determine. The stone tools of these primitive beings were cleverly made but were utilitarian in nature. The Neanderthals disappeared about 35 thousand years ago.

About 40 thousand years ago, anatomically modern humans suddenly began to behave in a radically new manner. They started to make a multitude of objects from bone and stone that had an artistic rather than a utilitarian purpose. Probably the oldest piece of fine art is an idealized sculpture of a horse carved on a mammoth tusk 32 thousand years ago. This and other pieces of prehistoric art demonstrate that these early humans were very similar intellectually to the humans of today.

What caused this radical modification of human behavior? There was no anatomical change observable in the skeletons. Only the behavior changed. *This rapid and radical behavioral change cannot be explained in terms of biological evolution.*

We will consider a new theory by Ian Tattersall, who is Curator of Anthropology at the American Museum of Natural History. Tattersall proposes that the development of sophisticated language produced this abrupt modification of human behavior. Is this the complete answer to the enigma? Even if Tattersall's postulate is correct, what caused this sudden development of sophisticated language? Was this mysterious event the result of divine intervention?

Even though the cells in our bodies are biologically related to those of other species, humans have characteristics that uniquely separate them from these species. The qualities that made us human appeared with remarkable suddenness about 40 thousand years ago. Why did this occur?

Chapter 2 summarizes the evolution of life on earth. The key stages of biological development are described. These give a logical thread to explain the primary evolutionary steps in the process that has produced the world that we observe today.

Our Sun and Our Milky Way Galaxy

Astronomers have discovered how our sun and other stars were born, how they live, and how they will eventually die. This information has been derived from astronomical observations, coupled with theoretical studies of the physical processes occurring within the stars. The enormous energy radiated from our sun is produced by the nuclear fusion of hydrogen to form helium, a process that is also implemented in the hydrogen bomb. Theoretical analyses of nuclear fusion have yielded detailed information of how our sun and other stars have evolved.

Studies show that our sun was created 5 billion years ago. A cloud of gas and dust was drawn together by the force of gravity. Eventually the pressure and temperature in the gas became sufficient to ignite nuclear fusion, and our sun was born. A disk of gas and dust orbited around the sun. From this disk, the planets were created, including the earth.

Recent evidence indicates that the formation of planets around a star is probably a normal development. A disk of gas and dust around a star appears to be essential requirement in the star's creation. Therefore there should be many stars having planets like earth that can support life.

The Mystery of Cosmology

The Hubble Expansion of the Universe

As we look beyond our Milky-Way galaxy, and examine our enormous universe with its billions of galaxies, we observe the strange phenomenon discovered by astronomer Edwin Hubble (1889-1953) in 1929. Our universe is expanding. Hubble discovered that, except for a few close galaxies, all galaxies are moving away from us with velocities approximately proportional to their distances. The average ratio of galaxy velocity divided by galaxy distance is called the Hubble constant. If we extrapolate galaxy motions backward in time, the whole universe seems to have emerged from a single point billions of years ago.

The radial velocity of a star or galaxy can be determined from its light spectrum. Radial velocity is the velocity in the radial direction, either toward us or away from us. A heated atom of a star radiates a light spectrum with a pattern of spectral lines that is unique for each kind of atom (for each element). This spectral pattern is shifted in wavelength by an amount that is approximately proportional to the radial velocity.

If a star is moving away from the earth, its spectral lines are shifted toward the red (toward longer wavelength), and if it is approaching the earth its lines are shifted toward the blue (toward shorter wavelength). A spectral shift toward the red is called a *redshift*, and a shift toward the blue is called a *blueshift*. The change of wavelength divided by the normal wavelength is approximately equal to the radial velocity divided by the speed of light. This wavelength shift is called the Doppler effect, which was named after Christian Doppler, who discovered it in 1842

Although the radial velocity of a galaxy can be easily measured from its spectrum, it is extremely difficult to determine the galaxy distance. Methods of measuring galaxy distance are explained in Chapter 4.

Hubble's estimates of galaxy distance were 8 times too small, and so his value for the Hubble constant was 8 times too large. If we use the latest average value for the Hubble constant, the whole universe appears to have emerged from a point in an enormous explosion about 15 billion years ago. This is the basic assumption of the *Big Bang theory*.

Serious Questions Concerning the Big Bang Theory

For many years, astronomers have been informing the public with absolute confidence that our universe was created with a "Big Bang" about 15 billion years ago. Nevertheless there is strong evidence conflicting with the Big Bang theory. When astronomers have extrapolated the universe backward to the instant of the Big Bang, they have run into severe contradictions. A multitude of artificial hypotheses have been proposed that attempt to explain the contradictions.

Our earth is 4.6 billion years old and our sun is 5 billion years old. An age of 15 billion years for our whole universe may superficially seem like a long time, but it is short in comparison to these ages, which we know with high certainty.

By studying the processes of stellar evolution, astronomers are able to estimate the ages of stars. As shown in *Scientific American*, May 2001 (p. 53), the most recent studies of stellar age have indicated that the oldest stars are no less than 11.5 and no greater than 14.5 billion years old. Yet many of the theoretical models of the Big Bang theory have yielded universe ages that are appreciably smaller than 15 billion years. How could the universe have evolved to achieve its present state within the short time allowed after the Big Bang?

Cosmic background radiation at microwave frequencies has been used by Big Bang proponents as proof that the Big Bang must have occurred. This radiation was discovered in 1965 and is claimed to be the cooled relic of optical radiation emitted from hot material about 300,000 years after the Big Bang. However, this radiation comes with extreme uniformity from all directions; its intensity varies with direction by only a few parts in 100,000. This shows that the universe must have been extremely uniform when this radiation was emitted.

But astronomers have found in recent years that the arrangement of galaxies within our universe is not at all uniform. Galaxies are concentrated in huge ribbon-shaped groups, and between these groups there are large spaces having few galaxies. In the very short age of the universe, how could this lumpy universe have evolved from the extremely uniform universe that presumably existed when the cosmic radiation was emitted? Instead of 15 billion years, a universe age of about 150 billion years seems to have been required.

Just a cursory examination of the evidence indicates that the dogmatic endorsement of the Big Bang theory is unwarranted. Scientists are usually very reluctant to regard a theory to be fact unless the evidence is overwhelming. Nevertheless, for many years astronomers have been treating the Big Bang theory as fact even though it is highly speculative. Because of this dogmatic approach, studies that contradict the Big Bang theory are discarded as being irresponsible.

Let us examine some other explanations for the Hubble expansion of our universe.

Alternate Explanations for the Hubble Expansion

The Hubble Expansion is Apparent

Some scientists argue that the universe expansion is only apparent. Effects other than velocity can cause a spectral redshift. These effects may make the universe appear to expand, even though it is not.

Paul Marmet [30] has proposed a redshift effect that has a solid scientific foundation, although it does not seem to be sufficient to explain the Hubble expansion quantitatively. He has shown that a photon of light loses energy when it collides with a gas molecule, and thereby experiences a redshift. The direction of the light does not change. This redshift effect cannot be observed in the earth's atmosphere, because the density of gas is too high. It cannot be observed in the laboratory, because it requires too long a path to create a measurable effect.

However, Marmet has shown that his effect explains the variation of the redshift of radiation across the disk of the sun. The redshift of radiation from the limb of the sun is greater than that from the center. Thus Marmet has experimental evidence to support his theory.

Details of the Marmet's theory are given in Appendix A. The theory indicates that every collision of a photon with a hydrogen molecule produces a redshift equivalent to a velocity change of 2 mm/sec. The number of collision is proportional to the density of the molecules and to the length of the path. As shown in Appendix A, the Marmet redshift would produce an apparent expansion of the universe that is consistent with recent measurements of the Hubble constant if the universe has an average gas density of 33,000 hydrogen atoms per cubic meter.

Gas densities that are very much larger than this have been observed by astronomers. In a gaseous nebula, the gas density is typically 100 billion hydrogen atoms per cubic meter.

Nevertheless, as shown in Appendices A and C, present estimates give an average density of matter in the universe that is very much smaller than the Marmet requirement. These estimates are based on gravitational effects associated with galaxy motions in a large galaxy cluster. Measurements of galaxy velocities within a cluster show that there must be about 400 times as much dark matter (which we cannot see) in the cluster as there is luminous matter (which we can see). Otherwise centrifugal force would make the cluster fly apart. This analysis yields an estimated average density of matter over the cluster equivalent to 3 hydrogen atoms per cubic meter.

The gas density required by the Marmet cosmology theory is 10,000 times greater than this estimate of 3 hydrogen atoms per cubic meter. This indicates that the Marmet redshift effect is probably not sufficient to explain the Hubble expansion of the universe.

On the other hand, Chapter 11 shows that the Marmet redshift effect gives a promising explanation for the extreme redshift of the quasar and the excess redshifts observed in certain galaxies.

The Steady-State Universe Theory

There is another well known alternative to the Big Bang theory, called the *Steady-State Universe* theory. This theory was proposed in

1948 by the noted astrophysicists Fred Hoyle (1915-2001), Hermann Bondi, and Thomas Gold. It was seriously considered by many astronomers until it was eclipsed in the stampede toward the Big Bang theory that occurred in the mid 1960's.

The Steady State Universe theory assumes that the age of the universe is infinite. The theory postulates that diffuse matter is being created throughout space to compensate for the Hubble expansion. The required creation of matter is far too small to be observed directly. Based on the present value of the Hubble constant, two hydrogen atoms would be created every year within a volume of one cubic kilometer.

This theory postulates that the diffuse matter gathers into huge clouds that coalesce to form new stars and galaxies. In this manner, our universe continually changes, and so it stays eternally young even though it is infinitely old.

What could produce this spontaneous creation of matter? The original Steady-State Universe theory did not have a clear answer. Is the matter created *out of nothing*? This may seem unacceptable, but should be no less acceptable than the Big Bang postulate that the whole universe was instantaneously created *out of nothing* at the Big Bang.

In 1965, strong microwave radiation was discovered coming from all directions, which was called *cosmic* microwave *background radiation*. This radiation had been predicted by Big Bang cosmologists as the cooled relic of optical radiation emitted 300,000 years after the Big Bang. The Steady State Universe theory did not have an immediate explanation for this radiation, and so the theory fell into disfavor.

This chapter describes a new variation of the Steady-State Universe theory, derived from the Yilmaz theory of gravity, which is a refinement of the Einstein theory. This new cosmology model predicts that the diffuse matter is created from energy that is radiated from stars. As shown in Appendix C, this new cosmology theory directly predicts cosmic microwave background radiation much more accurately than does the Big Bang theory.

Although support for the Steady State Universe theory declined rapidly in the mid 1960's, this theory was not disproved. The weaknesses of the original Steady-State Universe theory are certainly no more severe than those of the Big Bang theory.

Since there are viable alternatives to the Big Bang theory that give possible explanations for the Hubble expansion, there is certainly no justification for treating the Big Bang theory as fact. It is merely a theory. In our present state of cosmological ignorance, the honest scientific attitude would be to admit, "We do not know the answer".

The Singularity Concept

Scientists often use the concept of a *"singularity"* in implementing mathematical analyses of physical processes. When applied to matter, a "singularity" is a body that is compressed to zero size and so has an infinite density of matter. When normally used, a singularity is merely a mathematical idealization to simplify the analysis.

However, modern cosmologists use the *singularity* concept to represent physical reality. They literally mean a condition where a body has shrunk practically to zero size, and has a density of matter that is essentially infinite.

The January 2001 *Scientific American* was devoted primarily to a discussion of the Big Bang theory. The featured scientist for this issue was P. James E. Peebles, who was acclaimed to be the *"father of modern cosmology"* (p. 37). The Oct. 1994 *Scientific American* had presented an article by Peebles and others entitled "The Evolution of the Universe" (pps. 53-65) [28]. The article began with:

"At a particular instant, roughly 15 billion years ago, all of the matter and energy we can observe, concentrated in a region smaller than a dime, began to expand and cool at an incredibly rapid rate."

Our universe has about ten billion galaxies, each of which contains many billions of stars. Peebles, the *"father of modern cosmology"*, is claiming that our whole universe was initially compressed into a region *"smaller than a dime"* at the instant of the Big Bang. This *singularity* concept is widely endorsed by modern authorities on the Big Bang theory, although many are vague about the actual size of the initial Big Bang "singularity".

The claim that our universe began as a singularity strongly violates our laws of physics, and therefore gives serious reason to question the theory. There is certainly no justification for treating the Big Bang theory as fact, which astronomical literature has done for many years.

The Einstein General Theory of Relativity

The singularity concept evolved from studies of the *Einstein General theory of Relativity*. The Einstein theory was a tremendous advance in our understanding of physics, and consists of a large body of very sound physical principles. However, it is specified by the *Einstein*

10 *The Scientific Story of Creation*

gravitational field equation, which was developed by Einstein in an intuitive manner; it was not derived rigorously. Many scientists have questioned the adequacy of this equation, and have proposed alternatives to it. This book will show that the Einstein gravitational field equation does not give a rigorous specification for the principles of Relativity. In short, Einstein did not quite get the correct answer.

The Einstein gravitational field equation is extremely complicated. Consequently, it could only be applied to very simple physical models during Einstein's lifetime, and so Einstein did not appreciate the weakness of his theory. It was not until the mid 1960's, about a decade after Einstein's death, that computers became available that could apply the Einstein gravitational field equation to complex physical models.

The singularity concept first evolved from an effect derived from the Einstein theory that was later called a "black hole". Einstein strongly opposed the "black-hole" singularity, and insisted that *"singularities do not exist in physical reality"*. It was not until after Einstein's death that computer studies proved that the Einstein gravitational field equation does indeed imply a singularity.

Since the 1960's, hundreds of scientists throughout the world have devoted their careers to the task of implementing computer solutions of the Einstein gravitational field equation. With computers, that extremely complicated formula could be solved in a manner unheard of in Einstein's day. The awe of the Einstein name has given great prestige to the scientists engaged in this research. These computer studies of the Einstein theory form the basis for the extensive research on the Big Bang theory.

In order to explore the question of how our universe was created, we must first understand the Einstein General theory of Relativity. The reader may feel that this is way over his head. However, we will see that the principles of Relativity can be explained in a simple physical manner, which can be readily comprehended by the average reader.

The Yilmaz Refinement of the Einstein Theory

In the early 1950's, Huseyin Yilmaz was studying General Relativity as part of his PhD research at the Massachusetts Institute of Technology. He discovered that Einstein had made an unnecessary approximation in one of his General Relativity analyses. Yilmaz implemented this analysis exactly, using equations that Einstein had developed for Special Relativity. He found that this yielded an exact solution to the principles of Relativity. (This exact analysis is given in Appendix E.)

From this exact solution Yilmaz derived in a rigorous manner a different gravitational field equation, which is the foundation for the Yilmaz theory of gravity. The Yilmaz theory has applied the principles of the Einstein theory, and so is not a new theory; it is a refinement of the Einstein theory. The Yilmaz theory does not allow a singularity and so has proven that the principles of the Einstein theory are inconsistent with singularities.

Yilmaz sent his analysis to Einstein, but Einstein died before he could read it. In 1958, Yilmaz published the first paper on his theory in the prestigious *Physical Review*. Since then he has published numerous papers to extend his theory.

Because the Yilmaz theory has an exact solution, it is very much easier to use than the Einstein theory. The elaborate computer analyses that have been developed to apply the Einstein theory are not needed with the Yilmaz theory. If the Yilmaz theory should become widely accepted, these computer techniques would become obsolete. Besides, the Yilmaz theory has refuted the physically impossible singularities that have been derived from the Einstein theory. Therefore, it is not surprising that the Yilmaz theory is opposed by the army of Big Bang theorists that are engaged in computer studies of the Einstein theory.

Einstein's Rejection of the Singularity Concept

The concept that General Relativity predicts a singularity was first proposed in a 1939 paper, which claimed that a star with a mass-to-radius ratio a quarter million times greater than our sun would become what was later called a "black hole". The paper predicted that the star would collapse until it forms a singularity having an infinite density of matter. As will be shown in Chapter 3, Einstein absolutely rejected the black hole singularity. Except for this 1939 paper, no scientist proposed that General Relativity predicted a singularity while Einstein was alive.

During the 1940's and 1950's, the eminent nuclear physicist George Gamow was the primary spokesman for the Big Bang theory. Gamow postulated that the initial universe had the density of a neutron star, and consisted of tightly packed neutrons. It would have weighed 300 million tons per cubic centimeter. Gamow considered this to be the maximum density of matter that is physically possible.

With this density, our initial universe would have been the size of the orbit of the planet Mars. This was the Big Bang concept during Einstein's lifetime. The Big Bang singularity (with the universe squeezed to be "smaller the a dime") was not proposed until the 1960's, a decade after Einstein's death.

Except for the 1939 paper on the "black hole", no scientist claimed that the Einstein theory predicted either a black hole singularity or a Big Bang singularity while Einstein was alive. The singularity concept drastically conflicted with Einstein's philosophy that his theories must be consistent with observational evidence.

The Yilmaz Cosmology Model

In 1958 Huseyin Yilmaz published the first paper on his theory in the prestigious *Physical Review*. In this first paper he applied his gravitational theory to a simple cosmology model, which assumed that the universe has a constant average density of matter that does not change with time. *Yilmaz found to his surprise that his theory predicted an expanding universe, just as Hubble had observed. The Yilmaz analysis indicated that the Hubble expansion is a natural relativistic effect that is directly caused by gravity.*

How can gravity, which always causes masses to attract one another, force the universe to expand? Strange as it may seem, this effect evolved rigorously from the equations of the Yilmaz gravitational theory. Chapter 14 will give a simple physical explanation for why gravitational force should make the universe expand.

After presenting the first paper on his theory, Yilmaz discovered that cosmology can be very speculative, and so he ignored cosmological applications of his theory in subsequent papers. He has concentrated on mathematical extensions of his theory and experiments that can be performed within our solar system to test his theory.

The author is applying the powerful Yilmaz theory to cosmology. The analysis of the Yilmaz cosmology model was extended, and has yielded some remarkable results. The Yilmaz cosmology model predicts a universe that satisfies the principles of the Steady State Universe theory, but does not have the limitations that led to its abandonment. Details of this model are explained in Chapter 14.

The original Steady-State Universe theory could not explain the cosmic microwave background radiation, which was discovered in 1965. However, as shown in Appendix C, the Yilmaz cosmology model directly predicts cosmic microwave background radiation. This prediction agrees much more closely with experimental data than any prediction derived from the Big Bang theory.

The Genius of Albert Einstein

Albert Einstein has been widely proclaimed to be the greatest scientist of the twentieth century. However, very few of those who endorse this concept recognize what Einstein actually achieved. An important goal of this book is to explain the Einstein theory of Relativity in a simple manner that can be understood by the average reader.

This book will show that errors in the Einstein theory have resulted in cosmological predictions that strongly conflict with observational evidence and our laws of physics. Nevertheless, this weakness of the Einstein theory should not diminish our great respect for the Einstein genius.

The Einstein theory of Relativity took a bold and profound approach to explain some fundamental physical enigmas. The basic principles of General Relativity are very sound, even though Einstein did not achieve a rigorous solution to these principles with his gravitational field equation. The complexity of this formula was so great that its limitations only became apparent after Einstein's death, when powerful computers could be applied to it. From the Einstein equation, computer solutions have obtained physical predictions that Einstein certainly would have opposed, because they drastically conflict with physical laws.

The Yilmaz theory of gravity has derived a rigorous solution to the relativistic principles that Einstein established, and so is not a different theory; it is a refinement of the Einstein theory. The fact that Yilmaz has achieved this rigorous solution proves that Einstein's basic approach to Relativity was sound.

The Yilmaz theory eliminates the physically impossible singularity predictions that have been obtained from the Einstein gravitational field equation. Therefore the Yilmaz refinement of the Einstein theory has shown that the basic principles of the Einstein theory are inconsistent with singularities.

Our Scientific Story of Creation

Now let us return to our scientific story of Creation. We start by summarizing the research of paleontologists. From the painstaking examination of countless fossils they have derived remarkable understanding of how life has evolved on earth.

Chapter 2

The Creation of Life on Earth

The history of our earth and the evolution of its life have been deciphered from rocks and the fossils they contain. How old is our earth? The oldest earth rocks solidified 4.3 billion years ago. A meteorite fell in Arizona that is 4.5 billion years old, and so the earth and other planets should be at least that old.

Rocks obtained from the moon have yielded extensive information concerning the formation of the earth, because the moon and the earth have experienced similar meteorite impacts. Most meteorite craters on earth have been erased by weather and volcanoes. The moon has no weather, and much of its surface has been unaffected by volcanism

From studies of earth and moon rocks, scientists have concluded that the earth was formed as a molten body 4.6 billion years ago.

Early Life

Water gathered in the oceans, and it is there that life developed. The water came partly from volcanic hydrogen, oxygen, and steam, but meteorites may have delivered most of it to the earth. Although the crust of the earth solidified 4.3 billion years ago, studies of moon rocks and craters have shown that until 3.8 billion years ago the earth was bombarded so heavily by meteorites it was a very hostile environment.

Biologists now classify all life forms into three categories: *bacteria* and *archaea*, which are simple cells without nuclei, and *eukaryotes*, which are much larger and more complex cells that contain nuclei. All multi-celled organisms consist of eukaryote cells.

Archaea were recently discovered in volcanic thermal vents on the ocean floor, and typically live at temperatures close to the boiling point. They can derive nourishment by combining volcanic hydrogen and carbon dioxide to form methane and water, and they also feed on sulfur. Archaea were first thought to be bacteria, but DNA studies show they are radically different. Strange as it may seem, the DNA of eukaryotes is more similar to that of archaea than it is to bacteria.

The primary chemical building blocks of life are amino acids. The ocean-floor volcanic vents have the chemicals and energy that could allow the basic amino acids to be formed by inorganic means. Therefore many biologists now believe that archaea were the first organisms. Archaea could have flourished in the hostile environment 4.3 to 3.8 billion years ago, when the earth was heavily bombarded by meteorites.

The first clear evidence of life occurred 3.6 billion years ago in the form of structures called *stromotolites* produced by colonies of *cyanobacteria*. These bacteria (which are also called *blue-green algae*) use photosynthesis to derive energy from sunlight.

The early organisms were bacteria and archaea, which lack nuclei. It took nearly one billion years before complex eukaryote cells containing nuclei appeared in the fossil record, 2.7 billion years ago. Fossils of eukaryote cells can be identified because these cells are much larger than bacteria and archaea, and contain different biochemicals. [40]

Photosynthesis is the process operating in cyanobacteria, algae, and terrestrial plants that uses energy from sunlight to produce food. Carbon dioxide from the air and hydrogen from the water are combined to synthesize carbohydrates. Oxygen is released into the air as a byproduct. Photosynthesis is implemented by chlorophyll, which looks green because it absorbs red and blue-violet wavelengths. Terrestrial plants are usually green, but seaweed (also called algae) can have different colors, because other pigments mask the green chlorophyll color.

Photosynthesis was first performed at least 3.6 billion years ago by cyanobacteria, which are simple bacteria cells. Single-cell eukaryote *algae* (with nuclei) probably appeared 2.7 billion years ago. Multi-celled algae, which we call seaweed, appeared 1.8 billion years ago. [41]

Oxygen was produced by the photosynthesis performed by cyanobacteria, single-cell algae, and seaweed. The oxygen gathered in the atmosphere and finally reached sufficient concentration to support complex animal life.

The First Animals

Abundant fossils of multi-celled animals suddenly appeared in our oceans at the start of the *Cambrian* period, 545 million years ago. Within 5-10 million years an evolutionary explosion occurred, in which animals from essentially all animal types (phyla) have been identified.

Scientists have long been amazed by the extremely fast evolutionary pace at the start of the Cambrian. Part of the explanation lies in the 100-million year *Ediacara* period, which preceded the Cambrian. Ediacara fossils are scarce because the animals had soft bodies, which rarely produced fossils. The Ediacara fossils are mud casts of the soft bodies. Some Ediacara fossils have been identified as jellyfish, soft corals, earthworms, and mollusks. However many are strange; some look like small fluid-filled air mattresses, unlike any animal or plant today. [41, 42, 50]

New insight into the ultra-fast Cambrian evolution has come from very recent studies showing that the earth was hit by as many as four extreme ice ages in the period from 750 to 580 million years ago. The following theory is gaining acceptance. Since snow is white, it reflects most of the sunlight back into space, and tends to make the earth colder. If glaciers get large enough, the cooling process can feed on itself, and the whole earth can freeze. Geological evidence suggests that extreme ice ages occurred in which the average earth temperature dropped to -50 °C (-60 °F). The oceans froze to a depth of half a mile and stayed in this condition for 10 million years. [43]

Volcanoes are a rich source of carbon dioxide. During rain, the carbon dioxide in the air reacts with silicates and carbonates in rock to form soluble bicarbonate compounds and silicon dioxide. This process removes much of the carbon dioxide from the atmosphere. However there was no rain during these extreme ice ages, and so a large concentration of carbon dioxide from volcanoes could have accumulated in the atmosphere. It is believed that this produced a greenhouse effect that finally melted the frozen earth, and soon created a very hot earth. Temperatures should have risen until the average earth temperature reached 50 °C (120 °F). After that, the temperature should have gradually declined and a normal climate returned. [43]

The severe ice ages between 750 and 580 million years ago would have killed most of the life on earth. However marine life could have survived in regions around undersea volcanic vents. These vents may have melted chimneys through the ice sheets, leaving isolated ponds of clear ocean water. In these isolated pockets of life, different forms of animal life could have developed, which may have created the many phyla of animal life that became evident when the Cambrian began. [43]

This theory of the Precambrian ice ages is only a few years old. It will be studied and debated for many years before it can gain general acceptance. Nevertheless, it is clear that extreme climates occurred between 750 and 580 million years ago. These extreme climactic conditions probably had a strong influence on the evolutionary explosion of animal life that occurred at the start of the Cambrian period, 545 million years ago.

Development of Fishes

Soon after the Cambrian explosion, 545 million years ago, there appeared a specimen from our own phylum, the *chordates*, which consists primarily of *vertebrates* (animals with backbones). The early chordates were nearly brainless filter feeders, but had a notochord nerve trunk along the body that was the predecessor of the spinal cord. These primitive chordates filtered microscopic organisms from the water and mud, and looked superficially like worms. [41]

About 510 million years ago, the notochord nerve trunk evolved into a cartilaginous backbone, and the first vertebrate appeared. This vertebrate was a jawless fish, which was similar to the modern lamprey. Many of these jawless fishes were armored. [50] (p. 114)

The lamprey is a parasite that feeds off larger fish by grabbing onto the side of the fish with a mouth that acts like a suction cup. The lamprey has no jaw, but has teeth that cut into the flesh of the larger fish. Early jawless fishes were probably filter feeders like their primitive chordate ancestors. Nevertheless, like the lamprey, they were much more complicated physiologically than the primitive chordates, having eyes, ears, and well developed circulatory systems.

The next milestone in vertebrate development occurred about 425 million years ago, when the first fish with jaws appeared. The jaw allowed a fish to escape from a simple filter-feeder life to become a predator or browser. The vertebrate jaw is a key feature that led to the success of the vertebrates. [44, 50].

The first jawed fishes (called Placoderms) wore protective armor. These primitive fishes were probably much heavier than water, and spent most of the time on the bottom. By 360 million years ago, all of the armored jawless and jawed fishes had become extinct.

A new innovation in fish evolution occurred about 415 million years ago, when the first member of the shark class of fishes appeared. Sharks and rays have cartilaginous skeletons, which are much lighter than bone, and so the fish is only slightly heavier than the water it displaces. This allows a shark-like fish to stay at a fixed depth with only a small upward force, which is achieved by a slow forward motion.

The shark-like fish could leave the region near the bottom and travel freely at different depths. Since they could move quickly, they did not need armor to protect themselves from predators.

The next major innovation in fish development occurred about 400 million years ago with the development of the air bladder in *modern bony fishes (Osteichthyes)*. The air bladder evolved as a modification of the intestine. Since shark-like fishes do not have air bladders, it seems likely that primitive Placoderm fishes also lacked air bladders.

A *modern bony fish* extracts air from the water to inflate its air bladder. Although this fish is basically much heavier than the water it displaces, the air bladder provides neutral buoyancy. This allows the fish to remain motionless at a fixed depth and thereby enables it to travel freely at various depths. With this great advantage of the modern bony fish, the primitive armored fishes, with their sluggish existence near the bottom, were forced into extinction. Today all jawed fishes either have air bladders (modern bony fishes) or have cartilage skeletons (sharks and rays).

There are two groups of modern bony fishes: the ray-finned fishes and the lobe-finned fishes. Almost all of the bony fishes today are in the ray-finned group. The lobe-finned group consists only of the lung fish and the Coelacanth, which is so rare it was believed to be extinct until discovered in 1938. Nevertheless, the lobe-finned fish was extremely important 360 million years ago, because it evolved into the amphibian, which in turn led to reptiles, birds, mammals, and eventually to humans.

Amphibians Invade the Land

Amphibians, which now consist of frogs, toads, newts, and salamanders, evolved from fishes and moved onto the land about 360 million years ago. How did this remarkable move from water to land occur? To live on land an amphibian needed lungs and feet. How did it derive these in an evolutionary process that consisted of a series of small changes?

The air bladder, which was a critical innovation in the evolution of the fish, allowed fishes to make the next major step in vertebrate evolution, the development of the lung. We see this in the modern lung fish, which has gills to obtain oxygen from water and a lung to obtain oxygen from the air. The lung was derived by modifying the air bladder.

One may ask, "Why does a fish need to breathe air?" The answer to this question has only been discovered recently by paleontologists. There is strong evidence that amphibians evolved from fish that lived in swamps. [45, 46] There is very low oxygen content in swamp water, and so these fish needed to obtain oxygen by breathing air with their lungs. We do not encounter this phenomenon today, because oxygen-poor swamps are now populated by amphibians and reptiles, rather than by fish.

Since many swamps are shallow, the fish that lived in swamps often used their fins as feet to move through the swamps. Eventually the fins developed feet-like forms, with well-defined toes. It was originally thought that these *fish with feet* were true amphibians. However, recent studies have shown that the feet were too weak to have supported the body on land, and could only have been used in the water.

An even more surprising aspect of these *fish with feet* is that some had more than five toes. All land vertebrates are called tetrapods (which means they have four limbs), and each of these limbs has five functioning toes. More than five toes can sometimes occur, but they are not functioning members. Sometimes the limbs have been lost (as in snakes) or some toes have been lost. Nevertheless, the fundamental skeletal pattern of all land vertebrates has four limbs, each with five toes.

This indicates that the first amphibian must have had four limbs, each with five toes. One of the fossil *fish with feet*, which was originally classed as an amphibian, was found to have eight toes. This fossil was clearly not an amphibian, nor could it have been the ancestor of an amphibian.

The subclass of modern bony fishes that includes the coelacanth and the lung fish apparently led to the first amphibian. These fish have lobe fins with a structure that is similar to the four limbs of tetrapods.

Thus we conclude that relatives of the coelacanth and the lung fish developed the ability to live in oxygen-poor swamps. They had lungs to breathe air, and developed feet to move efficiently through shallow swamps. One of these, having five toes on each foot, developed feet that were strong enough for it to crawl onto land. This animal was the first amphibian.

Why did these amphibians move onto land about 360 million years ago? The obvious attraction was food. Plants had spread over the land 70 million years earlier. These plants were followed by millipedes and other bugs that fed on the plants. The first amphibians ate these bugs. It was only later that amphibians developed the ability to eat plants.

Spread of Plants over the Land

Let us backtrack and see how plants spread over the land. Without land plants, there could be no land animals. We have seen that multi-celled marine plants, which are called algae or seaweed, appeared about

1.8 billion years ago. Terrestrial plants evolved much later, because their physiology is more complicated. They require specialized structures to obtain and hold moisture and nutrients.

About 430 million years ago, primitive terrestrial plants moved from water onto land. These were related to *liverworts* and *mosses*. Liverworts, which are more primitive than mosses, are short leafy plants that sometimes resemble livers. Mosses and liverworts lack a vascular system for conducting fluids, and so are low plants that require abundant moisture. These plants reproduce by spores. Unlike a seed, which has many cells, a spore is a single cell.

Next came the *ferns, club-mosses,* and *horsetails,* which have vascular systems for conducting fluids. These plants also reproduce by spores. Modern club-mosses are often called ground pines, because they can look like miniature pine trees. However many ancient club-mosses resembled palm trees, having grass-like leaves on top of a long trunk.

Horsetails are hollow reeds, like bamboo stalks, that have leaves like pine needles growing from the joints between the reed segments. Modern horsetails are typically 2 to 6 feet tall, but some ancient species were 60 feet tall and one foot in diameter.

In the *Carboniferous* period (354-295 million years ago), giant ferns, club-mosses, and horsetails, 50 to 130 feet tall, formed great tropical jungles. These produced massive coal deposits that we now use for fuel.

Cycads and *conifers,* which have seeds, appeared about 300 million years ago. Conifers bear cones and consist of pine, spruce, cedar, fir, etc. Cycads resemble palm trees, but are biologically very different. Most cycads have become extinct, and only a few species survive today. Although cycads were originally very successful, they have been largely replaced by flowering plants.

Flowering plants evolved much later, in the middle of the dinosaur period. The oldest definite fossils of flowering plants are 120 million years old, but flowering plants probably appeared early in the Cretaceous period, which started 144 million years ago. Birds evolved about the same time as flowers. What a coincidence that *flowers* and *birds,* which so enrich our lives with colorful beauty, were created about the same time!

Flowering plants have the advantage over earlier plants that they reproduce rapidly. The noted dinosaur authority, Robert Bakker, has proposed that dinosaurs greatly aided the development of flowering plants, because they devoured vegetation so rapidly. This produced large open spaces where flowering plants could germinate and grow.

Fungi, which consist of mushrooms, molds, yeast, etc., were originally classified as plants. However fungi are radically different from plants that contain chlorophyll, and so are now placed in a separate kingdom. Like animals, fungi achieve nutrition by digesting carbohydrates, but fungi perform this digestion externally by excreting enzymes. Fungi reproduce by spores.

Lichens played a key role in the spread of terrestrial plants over the land. A lichen is a symbiotic combination of fungus and algae (or cyanobacteria). The fungus forms a body that gathers and holds water and nutrients, and the algae perform photosynthesis to produce carbohydrate food. Lichens can grow on bare rock. They gradually dissolve the rock, converting it into soil on which terrestrial plants can grow. The earliest definite fossils of lichens are 400 million years old. Lichens probably existed before that, but they live in conditions where fossilization is rare, and so the lichen fossil record is meager.

As explained earlier, carbon dioxide in rain converts the carbonates of rock into soluble compounds. This process breaks down rocks and thereby helps to form soil in which plants can grow. Glaciers grind up rocks and so are also important means of making soil.

Single-celled eukaryotes (cells with nuclei) are called *protists*, and first appeared 2.7 billion years ago. Protists consist of plant-like cells (such as diatoms), animal-like cells (such as protozoa and amoebae), and fungus-like cells. Plant-like protists are also called algae.

The Reign of the Reptiles

As stated earlier, the first amphibians moved onto the land about 360 million years ago. The first *reptile* appeared 25 million years later. Reptiles have amniotic eggs like bird eggs, which allow them to reproduce on land. Amphibians, such as frogs, have small fragile eggs (like fish eggs) that must hatch in water.

Many of these early amphibians were much larger than modern amphibians, some being 13 feet long. Amphibians dominated the land for 45 million years until they were displaced by reptiles.

Within 10 million years after the first reptile appeared, the reptiles separated into two major groups: (1) the *Synapsid reptiles*, which evolved into mammals, and (2) the *Diapsid reptiles*, which evolved into lizards, crocodiles, dinosaurs, and birds. It is often thought that a third group, the *Anapsids*, evolved into turtles, but turtles may have descended from Diapsids. We know that the Synapsid reptile was related to the mammal because, like the mammal, each side of the skull has a single hole behind the eye socket. In contrast, a Diapsid reptile skull normally has 2 holes behind the eye socket. [47]

The Synapsids

The Synapsid reptiles (the ancestors of mammals) quickly became the dominant reptiles. The Synapsids controlled the land for nearly 70 million years, from 315 million years ago until the end of the *Permian* period, 248 million years ago. At that time a mass extinction occurred. This was by far the most catastrophic extinction since the Cambrian began and eliminated 95 percent of all animal species. [47]

The early Synapsids were Pelycosaurs. A common example was the 10-foot long carnivore, Dimetrodon. It looked like a giant lizard with a sail-like fin on top of its body, which was probably used to control body temperature. Although an ancestor of mammals, Dimetrodon is often included in the "dinosaur" collection of children's toys. [47]

In the late Permian, the primary Synapsids were Therapsids, which, unlike Pelycosaurs, may have been warm blooded. Therapsids typically had lizard-like bodies but had teeth like mammals. Usually the body was held high above the ground, but the legs sprawled sideways like a crocodile. A large herbivore, Moschops, was 17 feet long and nearly the size of an elephant. Gorgonopsid was a vicious carnivore with large canine teeth like a saber-tooth tiger. A common herbivore, Dicynodont, had a turtle-like beak instead of teeth. [47]

Some of the Synapsid reptiles survived the Permian extinction, but in the following *Triassic* period the Synapsids were eclipsed by a group of Diapsid reptiles, called Archosaurs, which includes crocodiles and dinosaurs. A small mass extinction occurred 225 million years ago, and after that the first dinosaur appeared. A much greater mass extinction occurred 205 million years ago, at the end of the *Triassic*. [48]

The causes of the mass extinctions ending the Permian and Triassic periods (248 and 205 million years ago) are being studied. Manicouagan Lake in Quebec, Canada lies in a 100-km wide meteorite crater, formed about 210 million years ago. [49] This impact from an asteroid or comet may have caused the Triassic extinction 205 million years ago.

However, new studies relate extinctions at the ends of the Permian and Triassic periods primarily to enormous volcanic activity. [51] Volcanoes released high levels of carbon dioxide, which produced a greenhouse effect to cause severe global warming. The high temperatures could have devastated plant and animal life. High carbon dioxide levels in the ocean at the end of the Permian were augmented by radical changes in ocean currents due to continental drift.

Although the causes of the mass extinctions at the ends of the Permian and Triassic periods will be strongly debated for some time, the consequences of these extinction events are clear. They radically influenced the evolutionary development of dinosaurs and mammals.

The Dinosaurs

The dinosaur movie *Jurassic Park* has popularized the name of the *Jurassic* period, which followed the Triassic and began 205 million years ago. For 140 million years, in the Jurassic and the subsequent *Cretaceous* period, gigantic dinosaurs controlled the land. Our children know many of the dinosaur names. The largest were the house-size Brachiosaurus and Brontosaurus (also called Apatosaurus). The weirdest looking were probably Stegosaurus, with triangular bony plates along its back, and the tank-like Triceratops with three horns on its head. By far the most ferocious predator was Tyrannosaurus Rex, which appeared in the Cretaceous period.

For many years it was assumed that dinosaurs were cold-blooded, and therefore sluggish like our modern reptiles. However there is abundant evidence that they were warm-blooded, which would have allowed strenuous activity. The dinosaurs in the *Jurassic Park* movie were portrayed with this concept in mind. Dinosaur paleontologist Robert Bakker was the primary proponent of the concept that dinosaurs were warm blooded and very active.

During the dinosaur era, the air was dominated by strange flying reptiles called Pterosaurs or Pterodactyls, which were closely related to dinosaurs. Some were small, but one Pterodactyl had a wingspan of 40 to 50 feet. These flying reptiles were very efficient in flight, but were awkward on the ground. [47] The Pterosaur probably evolved about the same time as the dinosaur. However its fossils are scarce, because its hollow bones were fragile.

The first *bird* (Archaeopteryx) appeared in the middle of the dinosaur era, and apparently evolved from a dinosaur. The original function of feathers was probably for insulation, and this very effective feature was later adapted to provide flight. Birds had the advantage over Pterosaurs that they are very nimble on the ground. Consequently birds were probably much more efficient than Pterosaurs in small sizes.

During the dinosaur era, the seas contained many kinds of Diapsid reptiles. This included the Plesiosaur (7 to 47 feet long), which had an oval body with four paddle-like flippers, and a neck so long it looked like a sea serpent. It was probably very effective in catching fish. The

Ichthyosaurs looked like giant fish, some being 50 feet long. They were the dinosaur-era equivalent of our whale. [47]

The Slow Rise of the Mammals

Many of the Synapsid reptiles that survived the mid-Triassic extinction 225 million years ago were quite similar to mammals, and from these came the first true *mammal*, which appeared 215 million years ago. The mammals were small during the dinosaur era. Most were the size of a mouse and none was larger than a domestic cat. Mammals remained tiny for 150 million years until the dinosaurs were eliminated.

Two key features of a mammal are accurate tooth occlusion (so that it can chew effectively) and the ability to chew and breathe at the same time. [48] Although the mammals evolved from Synapsid reptiles, the early Synapsid reptiles, and many of the Synapsids that became extinct in the Permian and early Triassic periods, did not look very much like mammals.

Sixty-five million years ago, at the end of the Cretaceous period, the dinosaurs were suddenly driven into extinction. It is generally believed that this was caused by a meteorite (an asteroid or comet) about 10 kilometers in diameter that hit the earth near the Yucatan peninsula in Mexico. This caused a disaster that killed nearly all plant and animal life on land and in the seas. The dinosaurs, pterosaurs, and marine reptiles became extinct. However enough of the small mammals and birds survived to continue their species, along with a few cold-blooded representatives of the once-dominant reptiles.

Probably the primary reasons that mammals and birds survived the mass extinction are that many were small and could eat a variety of foods, like seeds and roots. The cold-blooded reptiles that survived the extinction had the advantage that a cold-blooded animal can live for long periods with little food. As was stated earlier, dinosaurs were probably warm-blooded.

Since 1978, Professor Dewey McLean of Virginia Polytechnic Institute [53, 54] has written extensively showing that enormous volcanic activity occurred at the end of the Cretaceous period. Volcanoes flooded a million square miles in India, and a pile of volcanic lava near Bombay is 1.5 miles thick. If volcanoes could have caused the massive extinction at the end of the Permian period, it seems clear that this volcanism should have strongly affected the dinosaur extinction. This suggests that two separate disasters may have contributed to the destruction of dinosaurs at the end of the Cretaceous period. Some paleontologists believe that many dinosaurs had become extinct before the meteorite impact.

With the competition from the dinosaurs eliminated 65 million years ago, the mammals took over the land and evolved rapidly. The birds flourished also, filling ecological niches in the air previously dominated by pterosaurs. Mammals and birds moved to the seas, as whales, porpoises, seals, and penguins, taking the place of extinct marine reptiles. The mass extinction 65 million years ago ended the *Age of Reptiles* and ushered in our modern *Age of Mammals*.

Many dinosaurs at the end of the Cretaceous period were probably more intelligent then the modern reptiles (turtles, lizards, snakes, and crocodiles} that survived the mass extinction. Nevertheless the mammals were apparently much more intelligent than the dinosaurs that they replaced. It seems likely that the 150 million years of dominance of dinosaurs over mammals resulted in a large increase in mammalian intelligence. Brainpower was probably a great asset in helping a tiny mammal from being eaten by a terrifying carnivorous dinosaur.

The *primate* order consists of humans, apes, monkeys, and a primitive group, called prosimians, which includes lemurs and tarsiers. Prosimians appeared 50 million years ago, monkeys appeared 35 million years ago, and apes appeared 25 million years ago.

Apes differ from monkeys in that they have no tail, they have a much larger brain, and they move in an upright manner when in the trees. Monkeys scamper on four legs along tree branches. Modern apes consist of the gibbon and orangutan, which live in Asia, and the chimpanzee and gorilla, which live in Africa.

The Ascent of Humans

Australopithecus and the Homo Genus

About 6 million years ago, an African ape, called *Australopithecus*, developed the ability to walk upright. Climate changes had thinned out jungle trees, and this ape needed to move over appreciable distances between trees. Walking upright was a natural evolution for an ape. It had the great advantage that the ape could carry clubs and rocks in its strong arms to defend itself against predators, as chimpanzees occasionally do today. The canine teeth were much smaller in Australopithecines than in chimpanzees. [55] (p. 105) This suggests that Australopithecines did not use their teeth for fighting as chimpanzees do.

Australopithecus had the same size brain as the chimpanzee, and, like the chimpanzee, ate primarily a vegetarian diet.

About 2.5 million years ago, the first member of our own genus, *Homo Habilis*, evolved from this upright ape. He had a much larger brain than Australopithecus and had the intelligence to make crude stone tools. Along with fruits and vegetables, he ate meat regularly, and the stone tools were used to butcher the animals. Eating animals was a great biological advantage, because it is much easier to derive nourishment from meat than from plants.

About 1.8 million years ago, *Homo Habilis* was replaced by *Homo Erectus*, who had an appreciably larger brain. Homo Erectus was tall (about 6 ft) and a good runner. Fossil remains of Homo Erectus are extensive, but those of Homo Habilis are limited. Consequently we have poor knowledge of the Homo Habilis skeleton.

Although the Australopithecine could walk upright, its abdomen was similar to that of the chimpanzee. It had a conical chest, a potbelly, and no waist. In contrast, the Homo Erectus skeleton was essentially the same as that of modern man, except for the head. The Homo Erectus skeleton permitted the flexible movement to allow fast running. The Australopithecines, with their chimpanzee-like abdomens, could walk but they could not run very well. [55] (p. 195)

The Australopithecines disappeared about one million years ago. This extinction may have been more the result of competition from baboons than from Homo Erectus. Since Australopithecus did not eat meat, it did not compete directly with Homo Erectus, but did compete directly with the newly emerging baboons.

Hunting Capabilities of Early Man

Some anthropologists have argued (with convincing success in their profession) that Homo Erectus could not have been an efficient hunter, and was probably an opportunistic scavenger. However, this concept seems doubtful.

All predators scavenge, including lions, and so we must assume that Homo Erectus also scavenged whenever he could. Nevertheless, scavenging alone could not have yielded a steady and reliable source of meat. Hyenas have the reputation of being scavengers, yet they are also effective hunters. The vulture lives exclusively by scavenging, but it has the advantage that it can search over an enormous area to find dead animals.

We must assume that Homo Erectus was a good hunter, because he had the following strong advantages in hunting:
(1) He had the intelligence to allow effective cooperative hunting;
(2) He was a fast runner;
(3) He could carry rocks and clubs in his strong arms to mount a deadly attack;
(4) He could throw rocks and clubs from a distance.

Let us examine these attributes.

Cooperative hunting. Wild dogs in Africa are very efficient predators because they cooperate when attacking an animal. With his much higher intelligence, Homo Erectus could have been a very effective cooperative hunter. For example, one Homo Erectus group could have herded animals slowly into an area where another group lay hidden in ambush. With a quick dash, the hidden group could have easily made a kill.

The chimpanzee provides a good example of the effectiveness of cooperative hunting. Chimpanzees catch monkeys by attacking as a group that surrounds a monkey in the trees. Because of this cooperative action, the chimpanzees can easily catch and kill their prey.

Wielding clubs and rocks. Chimpanzees sometimes throw clubs and other objects when being pressed by a predator. However, they are not very effective in this regard, because they cannot stand firmly on two legs. In the arms of Homo Erectus, clubs and rocks would have been formidable weapons.

Throwing rocks and clubs. By throwing an object from a distance, Homo Erectus could have stunned or knocked down an unwary animal. With a quick dash, the animal could have been subdued and killed. The great skill of a baseball pitcher today may well be the result of human physical capabilities that evolved over millions of years.

It is hard to explain why Homo Erectus developed a body that allowed efficient running if it was not used for hunting. He could not run fast enough to outrun a predator.

From Homo Erectus to Anatomically Modern Humans

Homo Habilis made crude pebble tools (called *Oldowan*) by striking one lava pebble against another. Much more complicated stone tools (called *Acheulian*) were made by *Homo Erectus*. [56]

Homo Erectus had the adaptability that allowed him to leave Africa and populate Europe and Asia. Homo Erectus was found in Indonesia soon after he appeared in Africa. He eventually learned to use fire. [55]

Starting about 600 thousand years ago, more intelligent beings evolved from *Homo Erectus*, which are known as *Archaic Homo Sapiens*. Their brains were larger, and the types of stone tools increased. Specimens appearing in Europe about 600 thousand years ago are also called *Heidelberg* man.

A more advanced stone tool technology (called *Mousterian*) appeared about 200 thousand years ago. *Neanderthal man* also appeared about that same time. He had a heavy frame and was physically much stronger than modern man. His head enclosed a brain as large as modern man, but with a different shape. He had a sloping forehead, a heavy brow ridge, a very large nose, protruding teeth, and a receding chin. He disappeared about 35 thousand years ago. [56]

The first anatomically modern human (*Homo Sapiens Sapiens*) appeared about 100 thousand years ago. The oldest fossils were found in the Middle East and in Africa. For about 60 thousand years, modern humans and Neanderthals coexisted. Modern humans and Neanderthals made essentially the same tools, which were utilitarian devices for butchering animals, for making hunting spears, etc.

Then, suddenly, about 40 thousand years ago, modern humans started making extensive artistic objects from bone and stone. These objects included, for example, necklaces, pendants, and bracelets. The earliest fine work of art known is a beautiful ivory carving of a horse made 32 thousand years ago from a mammoth tusk. Caves in Lascaux, France contain superb drawings of animals that are 17 thousand years old, and are accompanied by a wealth of abstract symbols. These works of art demonstrate that those who made them were as intelligent as humans are today.

What caused this sudden and drastic change in the behavior of anatomically modern humans 40 thousand years ago? This question is probably the greatest mystery in the evolution of humanity.

The Development of Sophisticated Language

Anthropologists have long considered language to be a key element for understanding the evolution of human intelligence. Apes cannot make articulate speech sounds, because their vocal tracts do not allow it. In humans, the vocal cords are deep in the throat. The large space above the vocal cords allows the formation of articulate speech.

The principle structures of the vocal tract are the larynx (which holds the vocal cords), the pharynx (the tube above the larynx, which opens into the oral and nasal cavities), and the tongue and lips. Basic sounds are generated in the vocal cords, which are modulated in the structures above. In apes and in newborn humans, the larynx is located high in the neck, and so the possible speech sounds are very limited. In adult humans, the larynx is located low in the neck and thereby allows a wide modulation of sounds. [39]

This characteristic of modern humans that permits articulate speech comes at a heavy price. Because the vocal cords are so low, adult humans cannot breathe and swallow at the same time. Consequently humans can choke to death when they eat. Apes can breath while they swallow, and so do not have this problem. Infant humans can breathe and swallow at the same time, because the larynx is high in the neck. This is essential to infants because it allows them to breathe while they nurse. [39]

The position of the vocal cords can be determined from fossils. Studies have shown that the vocal tracts of Homo Erectus were beginning to be modified about 2 million years ago to allow articulate speech. The larynx reached its present low position at least 600 thousand years ago, as shown in the skull of Heidelberg man. [39] Since this vocal tract modification prohibits simultaneous breathing and swallowing, it must have had a strong biological advantage. Therefore it is reasonable to assume that, from the beginning of our Homo genus, articulate speech was an important characteristic.

Ian Tattersall, Curator of Anthropology at the American Museum of Natural History in New York, has presented a new theory to explain the revolutionary change in behavior of anatomically modern humans 40 thousand years ago. This is explained in a December 2001 *Scientific American* article by Tattersall, "How we came to be human", and in his associated book [39]. This change was far too rapid to be explained by physical evolution, and so he concludes that it must have been a cultural change.

Tattersall believes that this new behavioral characteristic was caused by the development of sophisticated language. With a sophisticated language, complex thoughts could be communicated, and this resulted in a radical change of behavior. This language achievement was probably a cultural development that applied speech and mental capabilities already present in anatomically modern humans.

This theory of Tattersall becomes clearer when we examine the vast difference between crude speech and sophisticated language. There have been cases of children who have grown up isolated from other humans and so did not learn to talk until they were over 12. They were eager to learn to talk and rapidly developed an appreciable vocabulary of disjointed words. However, they could not connect these words together to form logical sentences. They had learned to speak too late in their lives to achieve a sophisticated language.

Chimpanzees and gorillas have been taught to communicate using sign language. Although they have learned a large collection of words, they cannot tie these words together into sentences.

Articulate speech would have been a strong advantage to even the early members of our Homo genus, because it would have greatly assisted the task of cooperative hunting. Communication with speech would have been a strong advantage in a hunting attack. Complex language was not required. Simple commands would have been adequate to coordinate a hunting operation.

It seems likely that members of the Homo genus up to Neanderthals and early anatomically modern humans could utter articulate sounds, which they used to make crude commands that consisted of separate words. This crude speech would have been adequate to satisfy the essential needs of primitive life.

The radical innovation that resulted in modern humanity may have been the development of a sophisticated language, which included the syntax principles for connecting words together into logical sentences. With a sophisticated language, complex thoughts could be conveyed, and this could have led to an enormous cultural expansion.

Neanderthals may have lacked the intellectual capability of achieving these new language skills, and so could not share in this cultural innovation. They did not have our high forebrain, which may be essential for a sophisticated language.

Sophisticated language developed by anatomically modern humans about 40 thousand years ago would have had practical as well as artistic advantages. It would have allowed them to compete more effectively against the Neanderthals for limited food. This may explain why the Neanderthals became extinct about 35 thousand years ago.

How the Homo Genus Evolved

Ever since Homo Erectus left Africa about 1.8 million years ago, the different groups of the Homo genus apparently stayed in sufficient contact with one another for the Homo genus to remain a single species This means that the different groups could and did interbreed. There is fossil evidence of limited interbreeding between Neanderthals and anatomically modern humans, and so these two groups apparently were of the same species.

We are concerned with an enormous interval of time in comparison to recorded history. The migration of Homo Erectus out of Africa may have been no faster than 25 miles within a 25 year lifetime. Over 1000 years this slow migration could have carried early man a distance of 1000 miles. When we consider the migrations of early man, we are not implying the rapid migrations that have occurred in recent times.

How did Homo Erectus gradually evolve to produce modern humans? Milford Wolpoff at the University of Michigan is the primary proponent of the *Multi-Regional Hypothesis*. It postulates that the slow migration of early humans, and the resultant interbreeding, caused humans to evolve together and thereby remain a single species. Advantageous genes were passed slowly from one group to another. Nevertheless, regional differences remained, and these differences form the basis for the racial distinctions that we see today.

The *Multi-Regional Hypothesis* is opposed by the *Replacement Hypothesis*, which postulates the replacement of one Homo group by another. Proponents of the Replacement Hypothesis maintain that modern humans evolved at least 100 thousand years ago (and maybe much earlier), probably in Africa. They migrated and eventually replaced all of the earlier populations.

Wolpoff argues against the Replacement Hypothesis, because anatomically modern man coexisted with Neanderthal man for at least 60 thousand years. During that period the two groups made the same stone tools. The perplexing relation between modern humans and the Neanderthal "cave men" has long been an issue of debate.

A possible explanation of this coexistence of two radically different types of humans is that it occurred in Europe during ice-age periods. The Neanderthals were apparently much better adapted to cold weather than anatomically modern man. The Neanderthals may have lived in the cold regions, while the anatomically modern humans lived in the warmer regions. If the two groups did not live in the same location at the same time, there would have been little contact between them. This assumption could explain why they remained biologically distinct.

Applying the Tattersall theory suggests that sophisticated language may have given the anatomically modern humans a great advantage over the Neanderthals about 40 thousand years ago. If the Neanderthals were incapable of achieving a sophisticated language, they could have lost in the competition for food. Modern humans may have learned the complex skills of making tight, warm clothing, and thereby could live in the cold regions populated by Neanderthals.

Outside Europe, with its ice ages, there may have been much more interbreeding, so that the populations were more homogeneous. Consequently all of the inhabitants in the other regions may have been able to learn the new complex language skills.

Here is an hypothesis of how modern humanity might have developed, which may explain the relationship between modern humans and the Neanderthal "cave men". Whether or not the reader accepts this explanation, it should at least help to illustrate the confusing aspects of this enigmatic relationship.

Was there a Divine Spark in the Development of Humanity?

Our explanation of the evolution of life on earth, from early microbes to the development of modern humanity, has applied the principles of natural selection. The natural selection process capitalized on random genetic mutations, augmenting those mutations that yielded advantageous characteristics. By this means the complex sophisticated life of today could have evolved from very simple life in a countless series of small changes that occurred over billions of years.

Is that all that there was? Was any divine purpose involved in this process? With DNA analysis one can demonstrate a remarkable uniformity in the biological structure of different species. It seems impossible to argue that the cells in the bodies of humans are not biologically related to the cells in the bodies of other species.

However, this does not prove that there was no divine spark in the development of humanity.

The revolution of modern human behavior that occurred about 40 thousand years ago is a phenomenon that cannot be attributed to biological evolution. Modern human behavior was created far too suddenly. We have attempted to answer this with the Tattersall theory, but even this does not explain why humans suddenly began to use sophisticated language.

Since modern human behavior developed too quickly to be explained by biological evolution, one can legitimately ask the question, "Was this development the result of Divine intervention?"

The Growth of Civilization

Regardless of the causes, modern humans suddenly began to display a very high level of intelligence about 40 thousand years ago. About 12 thousand years ago, this intelligence led to a revolution in human activity with the development of agriculture. Before agriculture, humans were limited to a hunter-gatherer way of life, which greatly restricted the population that could be supported by the land.

Agriculture began about 12 thousand years ago with the planting of grains and the domestication of sheep, goats and pigs. With agriculture, the population density could dramatically increase. About 10 thousand years ago, the first towns of appreciable size appeared. One of the two earliest known towns was at the site of the Biblical town of Jericho.

After the development of agriculture, complex civilizations evolved rapidly. The pyramids of Egypt are probably the most striking evidence of this. Those marvels of engineering were constructed nearly 5 thousand years ago. It is hard to believe that the pyramids of Egypt were as old to the Classical Greeks as those Greeks are to us.

We have traced the evolution of life on earth, and the progression of humanity, down to the dawn of historical civilization. There is healthy debate over the details of this story, but the general outline of the process seems to be broadly accepted.

Beyond the Earth

Now let us raise our eyes from the earth to the heavens. We leave the realm of the paleontologist and enter that of the astronomer. Like the paleontologists, the astronomers have derived a tremendous amount of information from painstaking observations. This has yielded remarkable understanding of the beautiful stars that shine in the sky, including that nearby star that we call our sun.

Geological Periods (millions of years ago):

Cambrian 545-490; *Ordovician* 490-443; *Silurian* 443-417; *Devonian* 417-354; *Carboniferous* 354-295; *Permian* 295-248; *Triassic* 248-205; *Jurassic* 205-144; *Cretaceous* 144-65; *Age of Mammals* 65-present. [50]

Major Periods of Vertebrates: Jawless fishes (Ordovician); early jawed fishes (Silurian); modern fishes (Devonian); amphibians (Carboniferous); Synapsids (Permian); first dinosaurs and mammals (Triassic); dinosaurs (Jurassic and Cretaceous).

Chapter 3

The Creation of Our Stars

Our Milky Way Galaxy

Nearly every summer I visit a rural location that is far from city lights. I am always excited to view the sky on a clear moonless night, to see countless stars shining in wondrous majesty. The most inspiring feature to me is that pale white pathway across the sky, called the Milky Way. The Milky Way encircles the celestial sphere, dividing it into nearly equal hemispheres. It is sad that so many people today have never seen our mystical Milky Way, because it is obscured by sky glow due to the reflection of electric lights from the atmosphere.

Our sun lies within an enormous spiral galaxy, which we call the Milky Way galaxy. This is similar to the Whirlpool galaxy, which was shown in Fig. 1-1 and on the front cover. It has the shape of a disk, which is 100,000 light years in diameter and 3500 light years thick at the center. Our sun is located about 2/3 of the distance from the center to the circumference. The sun lies within a spiral arm and is near the center of the galaxy disk. The Milky Way pathway across the sky is our edge-view of the galaxy disk. It is the light emitted by billions and billions of very distant stars that lie within our Milky Way galaxy.

Because the Milky Way galaxy is rotating, our sun is moving at a velocity of 250 kilometers per second. The galaxy makes a complete revolution in 200 million years. Our galaxy contains 100 billion stars, which have a mass equivalent to 10 billion suns. It seems hard to realize that the Milky Way pathway across the sky is our view of a galaxy so huge that light takes 100 thousand years to travel across it.

Stars in our Milky Way galaxy vary in size from about 1/10 of the mass of our sun to about 100 times the sun mass. If a star has less than

1/12 of the sun mass, it cannot achieve sufficient temperature to produce nuclear fusion, and so does not ignite. There may be many dark bodies, called brown dwarfs, that are too small to achieve nuclear fusion and become a star.

It is often stated that our sun is an *"average sized star"*. This is probably a good qualitative statement. However, the question of what one means by an *"average-size star"* is complicated, as we will now see.

An equation for the distribution of stellar masses was derived from astronomical observations in 1955 by Salpeter, and is given by Peacock [6] (p. 379, Eq. 12.85). This equation is expressed graphically in two different ways in Figs 3-1 and 3-2. Later studies have obtained somewhat different distributions, but the Salpeter formula should be adequate for our purpose.

Figure 3-1 shows the percentage of the total number of stars that have more than a specified mass, relative to the sun mass. This figure shows that 4.6 percent of the stars have masses greater than the sun, and so 95.4 percent of the stars have smaller masses. This figure also shows that the median star size is 1/6 of the sun mass. Fifty percent of the stars have greater than 1/6 of the sun mass, and fifty percent have less than this mass.

Figure 3-2 presents the data in a different manner. It shows the total amount of mass that is contained in the stars that have less than a specified mass ratio. This shows that about 60 percent of the total stellar mass consists of stars that are smaller than our sun, and so 40 percent of the total stellar mass is contained in stars larger than our sun.

Figure 3-2 shows that even though the median star mass is 1/6 of the sun mass, only 18 percent of the total stellar mass is contained in the stars smaller than the median star size.

Figure 3-1 shows that only 0.5 percent of the stars have masses that exceed five times the sun mass. Figure 3-2 shows that 17 percent of the total mass is contained in these very few stars that are larger than five times our sun mass.

Figure 3-2 shows that half of the stellar mass is contained in stars having less than 60 percent of the sun's mass, and half is contained in larger stars. Therefore it is reasonable to call our sun an "average-size star", but our sun is certainly not a star of "median size".

We will make further use of the data in Figs 3-1 and 3-2 as we proceed in our study of the stars.

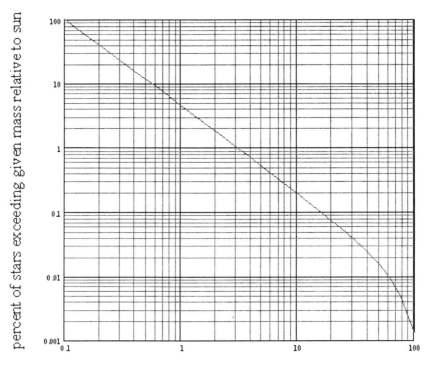

Figure 3-1: Percent of stars that exceed a given mass relative to the sun

Enormous Distances to the Stars

After our manned trip to the moon, the concept of space travel became very popular, and we began to experience through fiction, in television, movies, and books, an abundance of fanciful voyages throughout our universe. I am sure that few people who are delighted by these stories realize how infinitesimal our trip to the moon was when compared with a voyage to even the nearest star.

Distances to stars are commonly measured in light years, which is the distance that light travels in one year. Light travels so fast that to our senses it seems to move instantaneously between points on the earth. With satellite telephone communication we have direct experience that the speed of light is finite. Communication satellites are in synchronous orbits, located 26,000 miles from the center of the earth. Consequently it takes 0.15 sec for the communication signal to travel from earth to a satellite, and the round trip takes 0.3 sec. This results in an awkward pause in a satellite telephone conversation. If one speaker starts immediately after he hears the other stop, there is a gap of 0.6 seconds between each comment, because two trips to the satellite are involved.

You can see this effect on television. When a commentator talks to a reporter via satellite, the reporter is very slow in responding to a question from the commentator, because of this time delay.

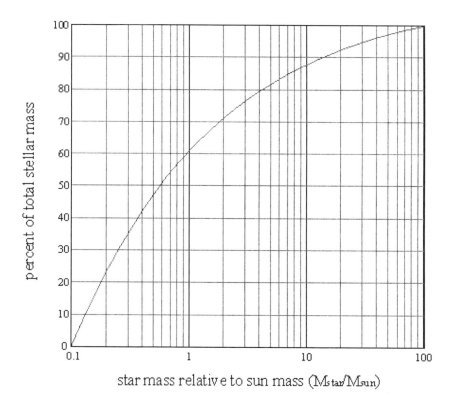

Figure 3-2: Percent of total stellar mass for stars having less than a given mass relative to the sun

The 1997 tests on Mars with the Sojourner Rover vehicle demonstrated a much more serious problem with the limitation of the speed of light. It took about 10 minutes for a signal to reach Mars, or 20 minutes for the round trip. Therefore if the Mars Rover vehicle observed an obstacle, the earth-based controller could not take corrective action until 20 minutes later. Consequently controlling the action of this robot on Mars was very difficult.

It takes only 1.4 seconds to transmit light to the moon, which is the farthest point to which man has traveled in space. Light from the sun takes 8.3 minutes to reach earth, and it takes 4.2 hours to reach Neptune. The time for light to travel from the sun to our outermost planet, Pluto, varies from 4.1 to 6.8 hours, because Pluto has a highly elliptical orbit.

These times are puny in comparison to the time for light to come from even the nearest of stars. It takes light 4.2 years to travel from the closest star, Proxima Centauri. A more worthwhile star to visit would be its much brighter neighbor, Alpha Centauri, which is 4.3 light years away and is about the size of our sun. There are 12 stars within a distance of 10 light years, 28 stars within 12 light years, and 62 stars within 17 light years. [25]

These results extrapolate to about 100 stars within 20 light years. In our region of the galaxy, the average distance between stars is 7 light years. The stars are packed much more closely in the nucleus at the center of our galaxy.

The nearest star, Proxima Centauri (4.2 light-years) is 100 million times farther than the moon (1.4 light-seconds), which is the longest space voyage yet made by man. We have sent unmanned space vehicles to observe the outermost planets of our solar system, and it took 12 years to reach Neptune. The nearest star is 9000 times further than Neptune.

Enormous as distances are to the nearby stars, they are extremely small in comparison to distances within our galaxy. Remember that it takes 100 thousand years for light to travel across the diameter of our Milky Way Galaxy. In the *Star-Trek* television series, the *Enterprise space ship* would zip around our galaxy with no trouble at all. Nevertheless light is so sluggish it could have traveled only 5 percent of the distance across our galaxy in the time that has elapsed since the pyramids were built by the Egyptians, 5000 years ago.

The closest large galaxy beyond the Milky Way is the great Andromeda M31 galaxy. This is a spiral galaxy that has twice the diameter of our Milky Way galaxy. The Andromeda M31 galaxy is 2.2 million light years away, a distance that is 22 times the huge diameter of our Milky Way galaxy.

The Whirlpool galaxy shown on the front cover, which we have used to illustrate our Milky-Way galaxy, is 35 million light years away. As astronomers look out into our universe, galaxies seem to extend to the observational limits of their telescopes. The Hydra galaxy cluster has been observed which is at a distance of 2.4 billion light years. This distance is 24,000 times the diameter of our Milky Way galaxy. *The size of our universe is so enormous it seems beyond comprehension.*

3. The Creation of Our Stars   39

The Search for Intelligent Extraterrestrial Life

When an unexplained phenomenon occurs on earth, it is often attributed to the work of aliens from outer space. When this explanation is doubted, a common reply is, "Are you so arrogant as to think that humans are the only intelligent beings in the universe?" Let us evaluate this comment.

Our space explorations have definitely shown that outside the earth, there is no other intelligent life in our solar system. We must look to a planet around another star. As we will see later, the formation of a solar system around a star appears to be a normal development. Let us consider the possibility of intelligent life on a planet orbiting a star within our Milky Way galaxy.

Our Milky Way galaxy has 100 billion stars. Figure 3-1 shows that 5 percent of the stars have masses between 0.7 times our solar mass and 1.4 times our solar mass. These may be good candidates for stars having planets similar to the earth. This gives 5 billion promising planets within our galaxy that might support life. We might expect that 10 percent of these may have advanced life, and 10 percent of the result may have intelligent life. Thus, we estimate that our galaxy may have 50 million planets having intelligent beings. If you think that this is too optimistic, throw in another factor of 5 as a safety factor. This still gives 10 million planets within our galaxy having intelligent life.

But what is the practical meaning of this estimate? Nearly all of these planets are so far away there is no possibility of our ever reaching them. We cannot communicate with them, and we cannot even make measurements to indicate that they exist. As far as we are concerned, their possible existence has no practical meaning.

If we are to consider intelligent extra-terrestrial life, we must limit our investigation to those stars within our galaxy that are close to earth. Even for these nearby stars, the distances are so vast, we might never be able to travel to the planets of these stars.

Nevertheless, let us approach this quest with optimism. As a minimum we eventually should be able to construct observatories in space with sufficient resolution to prove the existence of habitable planets around other stars. If these planets contain intelligent life, we may be able to communicate with them. A bit later we will take a closer look at nearby stars to determine which ones might possibly have intelligent life.

The Birth of Our Sun and Our Earth

Let us examine the process whereby our sun with its solar system, including the earth, was formed. Astronomers have learned a great deal about the birth of our sun by observing stars in various stages of evolution. These observations were combined with theoretical studies of the thermonuclear reactions that are occurring within the stars.

A star is formed when a thin cloud of gas and dust is drawn together by the force of gravity. As the cloud contracts, energy is released because of the gravitational contraction, and this energy heats the gas cloud. Eventually a temperature of about 10 million degrees Kelvin is reached at the center of the cloud. This temperature is sufficient to ignite nuclear fusion, and a star is formed.

The Kelvin temperature scale is measured relative to absolute zero temperature, which is at -273 degrees Celsius. At absolute zero temperature, the random motion of molecules is zero. The temperature in degrees Kelvin is obtained by adding 273 degrees to the temperature in degrees Celsius. A *degree Kelvin* is now officially called a *Kelvin*.

In the primary nuclear fusion process of a star, four hydrogen atoms are fused to form one helium atom. This process reduces the mass by about 0.7 percent, and an enormous amount of energy is released. Every gram of hydrogen that is converted into helium releases 180,000 kilowatt-hours of energy. One gram is 1/3 of the mass of a United States penny.

The heat generated within the star creates pressure that offsets the compressive force of gravity, and so the star reaches a stable condition. Our sun has been radiating nearly the same amount of energy since it was formed 5 billion years ago, and will continue to do this for another 5 billion years.

Angular momentum. To understand the formation of the solar system around the sun, we must consider the property of *angular momentum*. Figure skaters understand this issue very well, because they apply it when they exert the maneuver that allows them to spin rapidly.

The figure skater gives the body a twist from the skates, with the arms stretched out. Then the arms are drawn close to the body. Angular momentum must be conserved, and this effect requires that the body spin rapidly when the arms are held close to the body.

This principle is illustrated in Fig. 3-3. Two equal masses are held together at a constant distance. These masses are rotating about the center of gravity. Each mass is at a radial distance r from this center of rotation. The velocity of each mass is indicated by an arrow denoted V.

(These arrows are called "vectors".) The total mass is defined as M, and so the mass of each body is M/2. The angular momentum of these two rotating masses is:

Angular momentum = MVr

These two masses could represent the two arms of the skater. When the arms are drawn toward the body, the radial distance r decreases. Since angular momentum is conserved, the velocity V must increase as the radius decreases. Therefore bringing the arms close to the body makes the skater spin faster.

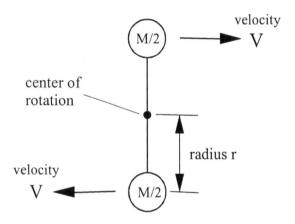

Figure 3-3: Pair of equal masses tied together and rotating about center of gravity; angular momentum is (M V r)

If the radius r is cut in half, the velocity V must double to keep the angular momentum constant. The frequency of rotation, which is the number of revolutions per second, is proportional to the ratio V/r. Therefore, when the radius r is cut in half, the velocity V is doubled, and the frequency of rotation increases by a factor of 4, because frequency is proportional to the V/r ratio. This effect causes the frequency of rotation of the skater to become very high when the arms are drawn tight to the body.

The sun experienced a similar effect as gravity caused the gas cloud to condense. The diffuse gas cloud was initially rotating because of the rotation of the whole galaxy. As the gas cloud condensed, it rotated faster and faster, because angular momentum was conserved. The centrifugal force due to this spinning would have stopped further collapse unless there were some way to eliminate angular momentum. In the early stages of the process, much of the initial angular momentum was lost by collisions of particles in the gas cloud. However, as the cloud became more compact, another process was needed to reduce angular momentum.

Our sun has more than 99.8 percent of the mass of the solar system, but has only 2 percent of the angular momentum. Jupiter has 70 percent of the angular momentum, and Saturn has nearly all of the rest. As our sun collapsed, angular momentum was transferred from the sun to a surrounding disk of gas and dust that became our solar system. Therefore a solar system around a star is probably a natural development. The solar system is needed to allow the star to lose its excess angular momentum, and thereby continue to collapse.

Plasma electric currents. How is angular momentum transferred from a star to its solar system? As reported by Lerner [18] (pp.188-189, 209), Nobel Prize winner Hannes Alfven (1908-1995), the pioneer in plasma physics research, has explained that the thin gas in space is ionized to form what is called *plasma*. The electrons are separated from the atom nuclei to create electric currents that flow in space. The current that flows through a square meter is very small, but the total current flowing over the vast reaches of space is huge. These electric currents produce magnetic fields, which generate enormous torques. Alfven concluded that the magnetic fields produced by plasma electric currents transferred angular momentum from the sun to its surrounding disk of gas and dust. This disk eventually coalesced to form our solar system.

The formation of our solar system. We are therefore led to the following explanation of how our sun and its solar system were created. As the cloud of gas and dust compressed, a disk of gas and dust remained around the condensing sun. Plasma electric currents flowed through this ionized gas to produce large magnetic fields. The magnetic field of the central sun rotated more rapidly than that of the surrounding disk. The interaction of these two magnetic fields created a torque that transferred angular momentum from the sun to the surrounding disk, thereby reducing the rotation rate of the sun. The effect is similar to the torque that is produced in an electric motor.

This shows that a surrounding disk of gas and dust appears to have been essential in the formation of our sun. The material in this disk eventually created our planets. Gravitational attraction caused particles in the disk to gather into grains, which grew into balls, which formed larger bodies. The bigger bodies attracted the smaller ones, until there were a limited number of planets orbiting the sun. Alfven has concluded that magnetic fields due to plasma electric currents would have accelerated the accumulation of gas and dust into planets.

The pressure of radiation from the sun pushed the gasses away from the sun. Consequently the inner planets (Mercury, Venus, Earth, and Mars) were formed primarily from dust particles, and so are solid. The outer planets (Jupiter, Saturn, Uranus, and Neptune) captured the gas molecules. The outer planets are very much larger than the inner planets, and consist primarily of gas. (We ignore the rogue planet Pluto, which is only the size of a moon.)

In this manner we can explain how our sun with its solar system was born. We should expect other stars to be created in a similar manner. This concept indicates that the formation of a solar system around a star is probably a normal development.

This conclusion is supported by recent astronomical observations of infrared stellar radiation. These observations show that stars less than 400 million years old are usually surrounded by dust clouds, but older stars are usually not. This suggests that planets formed around the older stars have swept up the dust clouds. [33]

The Life and Death of Our Sun and Similar Stars

Nuclear fusion and nuclear fission. Stars obtain energy by combining (or fusing) the nuclei of light atoms to form heavier atoms. This is the process that is implemented in the hydrogen bomb. The atomic bombs that were dropped on Japan in World War II used nuclear fission, in which the atoms of heavy radioactive atoms are split to form lighter atoms. Nuclear fission is also used in nuclear power plants.

In an atomic bomb, the radioactive decay process of heavy radioactive elements is accelerated to achieve a runaway condition. *Nuclear fission* splits the radioactive atoms to form lighter atoms and releases a large release of energy.

Nuclear fusion requires an extremely high temperature. This is achieved in the hydrogen bomb by using an atomic bomb to ignite hydrogen. This causes *nuclear fusion*, which converts the hydrogen into helium. *Nuclear fusion* releases much more energy per gram of material than does *nuclear fission*, and so a hydrogen bomb is very much more powerful than an atomic bomb.

For over 40 years, scientists have struggled to develop a controlled nuclear fusion process for generating power, but have been unable to achieve the extremely high temperature that is required. Although the center of the sun performs nuclear fusion at a temperature of 10 million degrees Kelvin, a nuclear fusion plant would require a temperature of 100 million degrees Kelvin, because the enormous pressure at the center of the sun cannot be produced.

If a controlled nuclear fusion process could be built, it would be a boon to mankind, because a nuclear fusion power plant would not release radioactivity and there is an enormous amount of fuel available.

Nuclear fusion in our sun. The nuclear fusion process occurring within our sun involves the fusion of four atoms of hydrogen to form one atom of helium. This process, which requires a temperature of 10 million degrees Kelvin, is occurring at the center of the sun. Our sun has performed this fusion for five billion years, and will continue for another five billion years. Near the end of this period the rate of energy release will greatly increase, and our sun will swell to become a red giant. Its surface will reach half way to the earth, and the earth will become so hot that all of its life will be destroyed.

After the hydrogen has been consumed, the nuclear fusion will stop, and the sun will shrink back to its present size. Gravity will force the sun to shrink further until the temperature at the center rises to 100 million degrees Kelvin. Another nuclear fusion process will then begin, in which three helium atoms are fused to form one carbon atom. The sun will continue in this state, radiating power in a stable manner for another billion years.

Contraction to a White Dwarf

When all of the helium of our sun has been converted into carbon, nuclear fusion will stop. Gravity will force the sun to contract until it shrinks to the size of the earth. During this contraction process, the sun will glow white hot from the energy released by gravity, and the sun will become a white dwarf star. Because of the small size, the energy radiated from the sun will be very small, and so the solar system planets will be extremely cold.

Today the average density of the sun is 1.4 times the density of water. The center of our sun has 100 times the density of water, whereas the surface has 1/1000 of the density of water, which is about the density of the earth's atmosphere.

When the sun contracts to its ultimate size, it will be about as large as the earth. Its density will be one million times that of water. One cubic centimeter of the sun will weigh one million grams, which is one thousand kilograms, or one metric ton. A metric ton is 2200 pounds, or slightly more than an English ton.

When the sun reaches this extremely high density, the contraction process will stop. The Pauli exclusion principle is a fundamental concept of quantum mechanics. This principle states that electrons can only occupy specific energy states. When our sun shrinks to the size of the earth, force from the electrons will stop the sun from collapsing further.

After the sun shrinks to the size of the earth, it will stop contracting and will slowly cool. This white dwarf star will gradually cool to become a black dwarf star, the highly compact dead ember of a once brilliant star.

There is abundant evidence showing that white dwarf stars (with their extremely high densities) actually exist. The brightest star in the sky is Sirius, which has a companion star that is a white dwarf. This white dwarf has been studied extensively. Its mass can be determined from its gravitational effect on the motion of Sirius. Sirius and its white-dwarf companion are 8.6 light years away.

Life Cycles of Other Stars

About 99.6 percent of all stars have life cycles similar to that of our sun. These stars fuse hydrogen to form helium, and then fuse helium to form carbon. Then they all shrink due to gravity to become white dwarf stars, which gradually fade into black dwarf stars. However, the rate at which a star proceeds through its life cycle varies greatly with its mass.

Most of the life of a star is involved in the fusion of hydrogen into helium. As shown in Ref. [26], (p. 250), the lifetime of a star in this state is given approximately by:

$$\{Star\ Lifetime\} = \{Sun\ Lifetime\}/(M_{star}/M_{sun})^3$$

The *Sun Lifetime* during this state is 10 billion years. The factor (M_{star}/M_{sun}) is the ratio of star mass M_{star} to the sun mass M_{sun}.

If the star has half the sun mass, this formula shows that the star lifetime is increased by a factor of 2^3, which is 8. If the star has twice the sun mass, its lifetime is decreased by a factor of 8. Thus, a star with half the sun mass has a lifetime of 80 billion years, while a star with twice the sun mass has a lifetime of only 1.25 billion years. The star lifetime represents the time spent in the initial nuclear fusion process, which converts hydrogen to helium. This is the major portion of the stellar lifetime.

During its nuclear fusion processes, a star converts mass into energy.

However, the relative loss of mass is very small, and so the mass of the resultant black dwarf star is nearly the same as that of the star when it was born.

The shorter the life of a star, the faster it radiates energy. Also the more mass that a star has, the more energy it has to radiate. This tells us that the rate of energy radiated from a star (the radiated power) is given approximately by

$$\{Star\ Power\} = \{Sun\ Power\}(M_{star}/M_{sun})^4$$

This shows that a star with twice the mass of our sun radiates 2^4 or 16 times as much power as the sun. A star with half the mass of our sun radiates 1/16 of the power of the sun.

The Meaning of Absolute Magnitude

To compare the powers radiated by different stars we must understand the meaning of *Absolute Magnitude*. Stellar *Magnitude* designates the light power **received** from a star; the dimmer the star, the greater is its magnitude. The unaided eye can observe stars with magnitudes up to 5 or 6. Magnitude gives a logarithmic measure of the received light power. This means that an equal magnitude difference represents an equal ratio of power.

A magnitude difference of 5 means a power ratio of exactly 100. A magnitude difference of 1.0 means a power ratio of 2.512, or approximately 2.5. A magnitude difference of 2 means a power ratio of approximately 2.5x2.5, which is 6.25.

Absolute magnitude designates the power that is **radiated** from a star. The *absolute magnitude* of a star is defined as the stellar magnitude that we would observe if the star were at a distance of 32.6 light years. The reason that astronomers use this distance is that it represents 10 *parsecs*, where one *parsec* is equal to 3.26 light years. The image of a star shifts over a year by an angular amount that is inversely proportional to its distance, because of the rotation of the earth around the sun. This shift is called *parallax*. One *parsec* is the theoretical distance of a star that would exhibit an annual parallax shift of ±1.0 arc second. The parsec is the primary distance measure used by astronomers.

The absolute magnitude of our sun is 4.8. Since a magnitude difference of 1 represents a power ratio of 2.5, a star radiating 2.5 times the power of our sun would have an absolute magnitude of 3.8 (i.e., 4.8

minus 1). A star radiating 1/2.5 or 0.4 times the power of our sun would have an absolute magnitude of 5.8 (4.8 plus 1).

Our Search for Stars with Intelligent Life

Let us use our knowledge of stellar characteristics to determine whether some nearby stars might have planets that could contain intelligent life. As shown in Ref. [32], astronomers are searching for planets around stars, but can only detect very large planets orbiting close to the star. The table on page 63 of Ref. [32] shows that many planets about the size of Jupiter have been detected in orbits smaller than the earth orbit. Small planets like our earth cannot be detected.

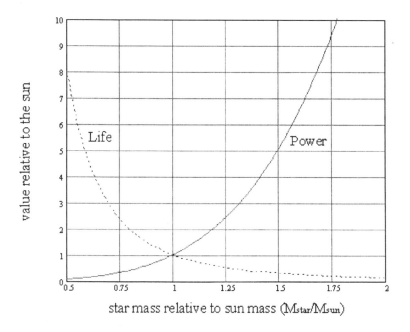

Figure 3-4: Effect of the mass of a star on the power and life of the star compared with the sun

Figure 3-4 gives plots of the relative power and the relative lifetime of a star versus its mass ratio with respect to the sun. A relative lifetime of unity represents 10 billion years, which is the lifetime of the sun.

Figure 3-5 presents the data from Fig 3-4 in a more convenient form. Relative power, relative lifetime, and relative mass are expressed in terms of the difference in absolute magnitude of a star with respect to our sun. A relative power of unity represents an absolute magnitude of 4.8, which is the absolute magnitude of our sun. Let us apply these plots to search for intelligent life.

We are picking candidate stars that may have planets with intelligent life. Our candidate star must radiate sufficient power to heat its planet to a temperature comparable to the earth. Let us assume that the minimum planet distance is the same as that of Mercury, which is about 40 percent of the distance of the earth from the sun. The intensity of solar radiation at Mercury is 6.25 times greater than that at earth. Therefore we want a star that radiates no less than 1/6.25 times the radiation from the sun. This represents a difference of 2 in absolute stellar magnitude.

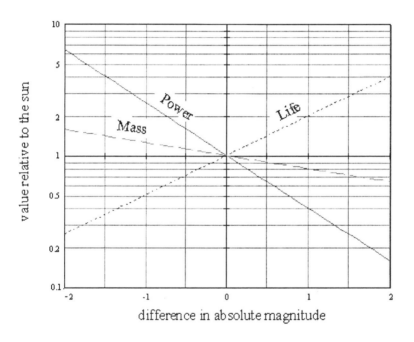

Figure 3-5: Variation of the relative power, life, and mass of a star with the difference in absolute magnitude of the star from that of the sun

Thus, our candidate star should ideally have an absolute magnitude that exceeds that of our sun by no more than 2.0. To take a more optimistic view, let us increase this to a difference of 3 in absolute magnitude.

Now let us consider lifetime. It took 5 billion years after our sun was formed before humans evolved on the earth. It took nearly 4.5 billion

3. The Creation of Our Stars 49

years before complex animal life appeared (at the start of the Cambrian period). Therefore it is reasonable to assume that it should take nearly 5 billion years for intelligent life to evolve in the planet of a candidate star.

The lifetime of our sun is 10 billion years. If a star has a lifetime less than half that of the sun (less than 5 billion years), it would not have had enough time to achieve intelligent life. Hence our candidate star should not have a lifetime less than half that of our sun. This translates to a factor of 2.5 in power, which is a difference of 1.0 in absolute magnitude. Our candidate star should not have an absolute magnitude that is more than 1.0 smaller than that of our sun.

Hence, we are looking for a star with an absolute magnitude that is no more than 1 less than our sun, and is no more than 3 greater than our sun. Since the absolute magnitude of the sun is 4.8, our candidate star should have an absolute magnitude in the range from 3.8 to 7.8.

The Dictionary of Astronomy [26] (p. 486) lists the 20 nearest stars, which are at distances out to 11.2 light years. Table 3-1 gives the five stars in this list that lie within our specified range of absolute magnitude. These are our nearby candidates for intelligent life.

Table 3-1: *Nearby candidate stars that might have planets with intelligent life*

star	distance (light years)	absolute magnitude
Alpha-Centauri A	4.3	4.4
Alpha-Centauri B	4.3	5.7
Epsilon-Eri	10.8	6.1
61 Cygnus A	11.1	7.6
Epsilon Ind	11.2	7.0

The two nearest stars form a binary star. They are separated by only 23 times the earth-orbit radius, and revolve around each other with an 80-year period. Gravitational interaction between the two stars may have adversely affected the formation of planets, and so it is difficult to predict the likelihood that these stars have planets like our earth.

This discussion has shown that there are stars reasonably close to earth that might have planets containing intelligent life. How will we ever reach these stars? I leave that as an exercise for the student.

The Supernova

The Dictionary of Astronomy [26] (p. 462, *white dwarf*) show that the life cycle we have discussed applies to stars with masses up to 8 solar masses. More massive stars follow life cycles that result in supernova explosions. Figure 3-1 shows that 0.4 percent of the stars exceed 8 times the solar mass. However, Goldsmith [24] (p. 191) states that only 0.1 percent of the stars have sufficient mass to become a supernova. Figure 3-1 shows that the Goldsmith condition corresponds to 16 times the solar mass.

Therefore we conclude that if the mass of the star exceeds somewhere between 8 and 16 times the solar mass, the star eventually explodes as a supernova. This occurs in somewhere between 0.1 and 0.4 percent of the stars. As shown in Fig 3-2, these stars contain between 7 and 13 percent of the total stellar mass.

Goldsmith describes the life cycle of a massive star in the following manner. This star proceeds through the primary process of fusing hydrogen to form helium. Like the lighter stars, it then proceeds to convert the helium into carbon. However, with its greater mass, the nuclear fusion process does not stop at this point. This massive star can achieve the temperature at its center that is required to form even heavier elements by fusion.

This massive star fuses carbon with the remaining helium to form oxygen; it fuses oxygen with helium to form neon; it fuses carbon with carbon to form silicon; and it eventually fuses silicon with silicon to form nickel, cobalt, and iron. Then the fusion stops, because the formation of an atom that is heavier than iron does not release energy; it absorbs energy.

These fusion processes proceed at faster and faster rates, because the successive fusion processes release less and less energy. It only takes a few weeks for the final stage that produces iron.

When fusion stops, massive gravitational collapse suddenly takes place. The force from electrons cannot limit the stellar density to that of the white dwarf, because the electrons are eliminated. The electrons are forced into the protons to form neutrons. This produces a star called a *neutron star* that consisting entirely of neutrons.

The Neutron Star

The *Pauli exclusion principle* still holds in the neutron star, but it now applies to neutrons. The principle states that two neutrons cannot occupy the same state (i.e., location). Material in the neutron star is squeezed until the neutrons are in direct contact The maximum density of matter allowed by the Pauli exclusion principle has been reached.

The density of a neutron star is enormous. It is 100 million times greater than that of the white dwarf, and so is 100 trillion times that of water. Suppose that we use earth-moving equipment to dig a hole that is 1500 ft deep over an area of 50 acres, to cover a square plot 1500 ft on a side. We put this material into a super-compactor, which compresses it into a volume of one cubic centimeter. That cube would have the density of a neutron star, which is 300 million tons per cubic centimeter.

When an electron is forced into a proton to form a neutron, a neutrino is released. This is an extremely small particle traveling at close to the speed of light. It normally passes freely through a body, and only rarely reacts with matter. Most of the neutrinos falling onto the earth pass through it as easily as light passes through our atmosphere.

Since the matter in the collapsing star is enormously dense, many of the neutrinos are absorbed by the outer layers of the star. This blows the star apart in an enormous explosion, and for a few months the star shines with the brightness of about 100 million suns. It is called a supernova.

The rate of supernova explosions in our Milky Way galaxy is about two or three per century. There are two major classes of supernovas. A second class of supernova apparently occurs when a white dwarf star is sucked into a massive star.

When the neutrinos impact with the surrounding material of the star, they cause nuclear reactions that produce atoms that are heavier than iron. The elements produced by the earlier fusion processes, along with these heavier elements produced by the neutrinos, are scattered as dust particles throughout the galaxy. These dust particles are gathered in the material that creates a new star.

It is the dust particles from earlier exploding supernovas that formed the solid matter from which our earth was built. Without supernova explosions, you and I would not be here.

Although only 0.4 (or 0.1) percent of stars are sufficiently massive to become supernovas, these stars are so large that the associated mass is appreciable. The stars have masses greater than 8 (or 16) times the mass of the sun. Figure 3-2 shows that 13 (or 7) percent of the total stellar mass occurs in these stars. Consequently, these exploding stars spread an appreciable amount of material throughout the galaxy.

The Pulsar

The neutron star was originally only a theoretical prediction until the pulsar was discovered in 1968. A pulsar emits a radio signal consisting of pulses at precisely timed intervals. When first observed, some people thought that these pulses might be radio signals from intelligent beings on a distant planet. However, studies soon indicated that they are probably generated by rapidly spinning neutron stars.

A pulsar has been found at the location of a supernova that exploded in 1054 AD, and was recorded by Chinese astronomers. This pulsar is surrounded by the Crab Nebula, which is a cloud of gas produced by the supernova. This pulsar generates precisely timed radio pulses at a rate of 30 times per second. It is slowing down at a rate that suggests an origin about 900 years ago. The fastest known pulsar generates 1000 pulses per second. (See Silk [21], p. 325.)

The theory indicates that a pulsar is a rapidly spinning neutron star. The magnetic field of the neutron star radiates radio beams from north and south magnetic poles along the magnetic axis. The energy creating this radio beam is derived from atoms being sucked onto the neutron star. Like the earth, the magnetic axis does not coincide with the spin axis, and so the radio beams spin in space as the neutron star rotates. When a radio beam passes the earth, we receive a pulse.

The pulsar star rotates at the pulse frequency. A pulsar with a frequency of 1000 pulses per second is a star rotating at 1000 revolutions per second. For a star to spin at the frequency of a pulsar, it must be extremely compact. The only explanation that satisfies the pulsar characteristics is a neutron star. Therefore, even though a neutron star has very strange properties, there is strong evidence that neutron stars actually exist.

The Black Hole

As explained by Goldsmith [24] (p. 193), the theory of neutron stars began in earnest under the guidance of J. Robert Oppenheimer, who later directed the Manhattan atomic bomb project. In 1939 he published a paper on neutron stars with the assistance of a graduate student, George Volkoff. This *Physical Review* paper was entitled "On Massive Neutron Stars". This gave the first firm prediction as to how massive a star must be in order to become a neutron star.

Also in 1939, Oppenheimer published another paper on highly dense stars. This *Physical Review* paper (Sept. 1939) with H. Snyder was titled

"On Continued Gravitational Contraction". [12] It was based on an analysis of the Einstein General theory of Relativity, and referred to the Schwartzschild solution to the Einstein theory.

When Einstein first developed the gravitational field equation for his General Theory of Relativity, he could only obtain approximate solutions to it, because the equation is so complicated. Karl Schwartzschild (1873-1916), who was cooperating with Einstein, was able to derive an exact solution by assuming a simple physical model of a star. Einstein published the Schwartzschild solution in the same year (1916) as the paper on his General theory. Unfortunately, Karl Schwartzschild died suddenly from disease even before his famous solution was printed. He had been a German army officer on the Russian front in World War I. The Schwartzschild solution formed the basis for experimental tests to verify the Einstein General theory of Relativity.

The Schwartzschild solution has a limit (called the *"Schwartzschild limit"*) when the ratio of mass-to-radius of a star reaches a value that is 240,000 times greater than that for our sun. When this limit is exceeded, the pressure inside the star becomes "imaginary", which indicates that the Schwartzschild analysis does not have an answer.

The Schwartzschild analysis assumed that the size of the star is constant. Oppenheimer and Snyder found that they could achieve a real solution from the Einstein gravitational field equation, when the *Schwartzschild limit* is exceeded, if they assumed that the diameter of the star decreases with time. Their paper concluded with:

"When all thermonuclear sources of energy are exhausted, a sufficiently heavy star will collapse. Unless fission due to rotation, the radiation of mass, or the blowing off of mass by radiation, reduce the star's mass to the order of that of the sun, this contraction will continue indefinitely."

In other words, this statement claimed that (according to the Einstein theory) a very massive star must collapse when the *Schwartzschild limit* is exceeded, and will shrink until it becomes a *"singularity"* having an *infinite density of matter*.

The next month Einstein responded to this paper with an extensive analysis in *Annals of Mathematics* [13], but the Einstein paper politely did not specifically refer to the Oppenheimer-Snyder article. Einstein concluded with

> *"The essential result of this investigation is a clear understanding as to why the 'Schwartzschild singularities' do not exist in physical reality. Although the theory here treats clusters whose particles move along circular paths, it does not seem to be subject to reasonable doubt that more general cases will have analogous results. The 'Schwartzschild singularity' does not appear for the reason that matter cannot be concentrated arbitrarily. And this is due to the fact that otherwise the constituting particles would reach the velocity of light."*

A few years later, Oppenheimer became the manager of the Manhattan atomic bomb project and never pursued this issue further. Neither did any other scientist while Einstein was alive.

In his rebuttal of the Oppenheimer-Snyder paper, Einstein insisted that the *singularity* condition derived from his theory is non-physical. The proposition that a star could contract indefinitely to form a *singularity* of infinite mass density severely violates our laws of physics. Einstein never accepted the physically impossible *singularity* concept.

The thinking of Albert Einstein on such issues is clearly explained in a recent and very thorough biography of Einstein, originally written in German by Folsing [23], who states (p. 381):

> *"Some of Einstein's admirers were tempted to see the general theory of relativity as a triumph of speculation over empiricism. This kind of misunderstanding made Einstein 'downright angry' [who said] 'This development teaches us something entirely different, indeed almost the opposite, namely that a theory, in order to merit confidence, must be based on generalizeable facts'. . . . To Einstein, facts were not only the starting point of his theory but also the keynote of any test of it."*

Although the concept that Oppenheimer and Snyder presented was not pursued officially while Einstein was alive, it did not die. When the mass-to-radius ratio of a star exceeds the Schwartzschild limit, the theory indicates that the star should be surrounded by a spherical surface called an *"event horizon"*, where the speed of light should be zero. At the event horizon, the ratio of mass-to-radius is equal to the Schwartzschild limit. Light theoretically cannot escape from within the event horizon, and so the star became known as a *"black hole"*.

The *black hole* became a popular theme for science fiction, and the public became well aware of the concept. There were many descriptions of the process of *"falling into a black hole"*. Despite the wide popular knowledge of the black hole, few people understand what must happen to the black-hole star that lies within the event horizon. They do not realize that the black hole star must theoretically collapse "indefinitely" until it shrinks to form a "singularity" having zero diameter and an infinite density of matter.

Some of the science fiction writers who considered the black hole notion have recognized that very strange conditions should exist inside the event horizon. Since light can only fall toward the center of a black hole, two adjacent atoms of a body could not interact in a physically meaningful manner. Energy could be transmitted from the upper atom to the lower atom, but not in the reverse direction. How then can real matter exist inside the event horizon?

About a decade after Einstein's death, powerful computers became widely available, and many scientists began to apply them to the Einstein theory. These computer studies apparently proved that Oppenheimer and Snyder had been right. Contrary to Einstein's objection, the computer studies found that the "Schwartzschild singularity", which is associated with the "black hole", is a required prediction of the Einstein gravitational field equation.

Therefore the black hole concept became widely accepted. In recent years a number of astronomers have made observations that are considered to be proof that black holes actually exist.

But how does one prove the existence of a black hole? A black hole does not emit any radiation to indicate its presence. What astronomers are apparently observing are the gravitational effects of massive, compact bodies. According to the Einstein theory, such bodies must be black holes. However, if we apply the Yilmaz theory, these bodies could also be massive neutron stars. The Einstein theory does not allow a massive neutron star, because it predicts that a massive neutron star must collapse to become a black hole.

If a star with 8 times the mass of our sun should collapse to become a neutron star, it would have a mass-to-radius ratio that reaches the Schwartzschild limit. Therefore, the Einstein theory predicts that the star must become a black hole. According to the Einstein theory, this neutron star should contract "indefinitely" until is becomes a singularity, having zero size and an infinite density of matter.

In contrast, the Yilmaz theory predicts that the neutron star does not contract any further. It remains a neutron star.

According to the Pauli exclusion principle, a neutron star has reached the maximum possible density of matter. Since the neutrons are tightly packed, there is no space left for further contraction. The Yilmaz theory agrees with this principle, and Einstein implied it in his rebuttal to the paper by Oppenheimer and Snyder.

Yet the widely accepted black hole concept predicts that matter can be compressed to a state that is many, many times more compact than a neutron star. In a black hole, the Pauli exclusion principle is radically violated. Matter as we know it ceases to exist. Protons, neutrons, and electrons are destroyed. Although the black hole was derived from the Einstein General theory of Relativity, it drastically conflicts with the philosophy that Einstein proclaimed throughout his lifetime.

Do black holes actually exist? The Yilmaz theory predicts that a black hole cannot exist. It is physically impossible.

Radiation from an Ideal Blackbody

A body at room temperature continually radiates energy in terms of heat. It also absorbs heat from the environment. Heat radiation is like light, except that it has a longer wavelength. The blacker the surface of an object, the better the object absorbs radiation, and the better it radiates. A *blackbody* is a physical idealization that radiates the maximum possible energy from a body at a particular temperature. If one knows the temperature of an ideal *blackbody*, one knows the spectrum and intensity of the radiation. Intensity is the power that is radiated from a body per unit of surface area.

One can build a nearly ideal *blackbody* by machining a spherical cavity inside a block of metal, and cutting a small hole into that cavity. Inside the cavity, radiation is emitted from each portion of the spherical surface, and continually reflects off other surfaces, until a small amount escapes out of the hole. The energy escaping from the hole is close to ideal *blackbody* radiation. The intensity of the *blackbody* radiation from the hole is always greater than the intensity radiated from the outer surface of the metal block. A *blackbody* is an idealized physical concept, yet it can be closely approximated by physical equipment.

The general spectrum of *blackbody* radiation is shown in Fig. 3-6. Diagram (a) shows the spectrum versus frequency and (b) shows the spectrum versus wavelength. The wavelength scale is expressed in terms of the half-power wavelength λ_h, and the frequency scale is expressed in terms of the equivalent half-power frequency f_h, which is equal to c/λ_h. Half of the power falls at wavelengths greater than λ_h and half falls at wavelengths less than λ_h. Area under the curve represents power, and so the areas under the curve on each side of the vertical line are equal.

3. *The Creation of Our Stars* 57

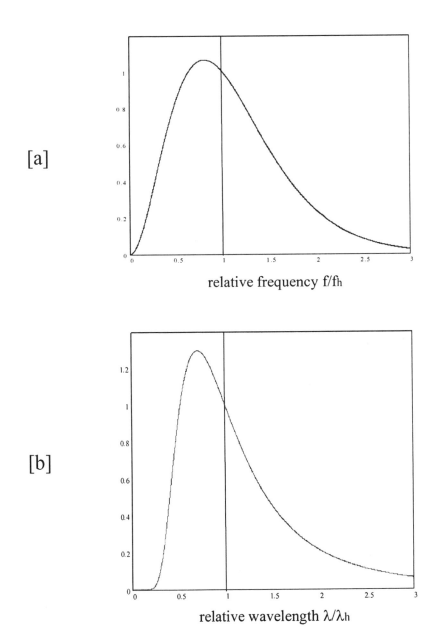

Figure 3-6: Normalized power spectra of a blackbody radiator; [a] versus frequency; [b] versus wavelength

The Greek letter λ (lambda) is used to denote wavelength. Since λ is the Greek equivalent of the Roman letter l (el), λ is a convenient symbol for denoting *waveLength*.

The temperature of a blackbody determines the actual frequencies that it radiates, but the shape of the spectrum is the same for all temperatures. The half-power wavelength λ_h is related as follows to the blackbody temperature T, expressed in degrees Kelvin (°K):

λ_h = 4.107/T millimeter (mm)

Kelvin temperature is measured above absolute zero temperature, which is -273 degrees Celsius. At absolute zero temperature, the random motion of molecules is zero. The temperature in degrees Kelvin is obtained by adding 273 degrees to the temperature in degrees Celsius.

The light radiated from our sun has a spectrum approximating that of an ideal blackbody at a temperature of 5770 °K. The corresponding value for the wavelength λ_h computed from the above formula is 0.000712 mm, or 0.712 micrometers (millionths of a meter).

The intensity of the radiation from a blackbody is related as follows to the temperature T of the body, expressed in degrees Kelvin:

P/A = 5.68 $(T/1000)^4$ watt/cm^2

Intensity is the power P radiated per unit of surface area A. For our sun, with a blackbody temperature of 5770 °K, this formula gives an intensity of 6300 watts per square centimeter at the surface of the sun.

The radiation from most stars approximates that of a blackbody. Consequently, by using the spectral curves of Fig 3-6 one can find from the spectrum of a star its half-power wavelength λ_h. From this one can calculate with the above equations the blackbody temperature T and the power radiated from the star per unit of surface area (the radiation intensity). If the distance to the star is known, one can determine the absolute magnitude of the star, and the power radiated from the star. Dividing the total power by the radiation intensity gives the surface area of the star, and from this one can obtain the star diameter.

These spectral curves of a blackbody are helpful in estimating the characteristics of stars. We will also use them in our discussion of cosmic microwave radiation. When this radiation was discovered in 1965, it was claimed to be proof of the Big Bang theory.

Chapter 4

The Creation of Our Universe

The View of our Universe in 1900

The 1929 Hubble discovery that our universe is expanding was the culmination of an extremely rapid growth in astronomical knowledge in the early part of the 20th century. This point is brought into focus by the following excerpt from a popular 1905 book on astronomy by Agnes Clerke, entitled *The System of the Stars* [20] (p.1):

"The question of whether nebulae are external galaxies hardly any longer needs discussion. It has been answered by the progress of research. No competent thinker, with the whole of the available evidence before him, can now, it is safe to say, maintain any singular nebula to be a star system of co-ordinate rank with the Milky Way. A practical certainty has been attained that the entire contents, stellar and nebula, of the sphere belong to one mighty aggregation, and stand in ordered mutual relations within the limits of one all-embracing scheme."

The term "nebula" had been used by astronomers since the 1700's to describe astronomical objects with extended images that do not move like planets or comets. In contrast, stars appear as points of light to a telescope. We now know that there are radically different kinds of nebulae, which consist of gaseous clouds, stellar clusters, and galaxies. However, even as late as 1905, it was generally believed that none of the nebulae represented distant galaxies. In the above quotation, Clerke was insisting that all nebulae lie within our Milky Way galaxy.

Soon after the Clerke book was published, astronomers were able to measure astronomical distances reliably, and thereby proved that our Milky Way galaxy is only one out of billions of galaxies that comprise our universe. Let us now look outside our Milky Way Galaxy to the vast universe that lies beyond.

The Local Group

We first observe our nearby companions, which are other members of our *Local Group*. This is a small cluster of galaxies, including our Milky Way galaxy, which are bound together gravitationally. The most impressive member of our Local Group is the great M31 spiral galaxy in the constellation Andromeda, which is 2.3 million light years away. It has a diameter of 200 thousand light years, twice the diameter of our Milky Way galaxy. It does not display so dramatic a shape as the M51 Whirlpool galaxy because we are observing it from nearly an edge view.

The next largest member of our Local Group is the M33 galaxy in the constellation Triangulum, which is 2.6 million light years away. In the southern hemisphere one can see the Large and Small Magellanic Clouds, which are 170,000 and 300,000 light years away. These are large clusters of stars. There are other smaller members of our Local Group. The most distant member is 5 million light years away.

The names M31, M33, and M51, which we are using to specify the different galaxies, are designations given in Messier's *Catalogue of Nebulous Objects*, which Charles Messier (1730-1817) published in 1784. Stars look like points of light to a telescope, but "nebulae" have extended images. Today we know that nebulae consist of galaxies, stellar clusters, and gaseous clouds. Messier developed his catalogue because he was looking for comets, and wanted to distinguish comets from fixed bodies. He discovered at least 15 comets. However, this catalogue is what made the Messier name famous, and it serves as the primary reference for locating extended objects in the sky. Messier lost his astronomer job when the French Revolution erupted in 1789.

The Hubble Expansion of Our Universe

As we look beyond our Local group we begin to experience the effect of the universe expansion, which Hubble discovered. Let us now consider the Hubble expansion in detail

Concept of the Hubble expansion

As was stated earlier, the velocity of a star or galaxy in the radial direction can be determined from its spectrum. If a star is moving toward us, its spectral lines are shifted toward the blue (they have blueshift),

4. The Creation of Our Universe

and if the star is moving away, the spectral lines are shifted toward the red (they have redshift). When we observe galaxies significantly beyond our Local Group of Galaxies, they all display redshift, which means that they are all moving away from us.

When astronomer Edwin Hubble compared the redshift of a galaxy with his estimate of galaxy distance, he discovered in 1929 that the redshift (and hence the radial velocity) of a galaxy is approximately proportional to its distance. This indicated that our universe is expanding at a nearly constant rate.

The concept of an expanding universe can be illustrated by assuming that a rubber band is being stretched at a constant rate. At an instant of time (called t_1) mark a series of dots on the band separated from one another by 10 mm, and label these dots as follows:

```
D'   C'   B'   A    B    C    D    E
*    *    *    *    *    *    *    *
```

The distance from A to B is 10 mm, from A to C is 20 mm, from A to D is 30 mm, etc. Now look at the band later (at time t_2) when it has stretched by 10 percent, so that the distance between neighboring dots has increased to 11 mm. The distance from A to B is now 11 mm; from A to C is 22 mm; from A to D is 33 mm, etc. Between time t_1 and time t_2, point B moves 1 mm away from A, point C moves 2 mm away from A, point D moves 3 mm away from A, etc. The relative velocity between points along the band is proportional to the distance between the points. This indicates that the rubber band is being stretched at a constant rate.

We can extend this analogy to two dimensions by considering a balloon that is being blown up at a constant rate. Mark on the surface of the balloon an array of dots separated by equal distances. As the balloon expands, the dots recede from one another at velocities proportional to the distances between the dots.

In 1929, Edwin Hubble (1889-1953) reported his discovery that the universe is expanding at a constant rate. This concept is called the Hubble law and the rate of expansion is called the Hubble constant, denoted H_0. The early measurements by Hubble had considerable error and variability. Greatly improved measurements have been made in recent years, assisted in particular by data from the Hubble Space Telescope. The average of recent studies gives a Hubble constant of about 20 kilometers/sec per million light years. Thus a galaxy at a distance of 10 million light years moves away from us (recedes) at a velocity of approximately 200 km/sec.

Our sun revolves along with our galaxy at a speed of 250 km/sec, and the earth revolves around the sun at a speed of 30 km/sec. Corrections are made for these velocities (and the relative motion of the sun within our galaxy) in determining the Hubble constant.

The Apparent Age of the Universe

The most obvious explanation for the Hubble expansion is that the universe began as a very compact body, which exploded billions of years ago to produce the universe of today. This is the basic concept of the Big Bang theory.

If we assume that the universe is expanding exactly at a constant rate, and has always done so, we can extrapolate the galaxy expansion in a simple manner backward in time. This assumption yields what the author calls the *Apparent Age of the Universe*, which we denote T_0.

The apparent age of the universe can be computed as follows from the Hubble constant. If the universe expands at 20 km/sec per million light years of distance, it expands at a velocity of 20,000 km/sec at a distance of one billion light years, and it expands at a velocity of 300,000 km/sec at a distance of 15 billion light years. The speed of light is 300,000 km/sec.

Thus, a Hubble constant H_0 of 20 km/sec per million light years implies that the universe expands at the speed of light at a distance of 15 billion light years. This suggests that the universe began as a point 15 billion years ago, and that the *Apparent Age of the Universe* T_0 is 15 billion years.

If the universe has always expanded uniformly at a constant rate, the apparent age of the universe T_0 would be the true age of the universe since the Big Bang. However studies based on the Einstein theory have predicted that the universe expansion rate should have been greater in the past, and so the true age of the universe should be less than the apparent age. Most *Big Bang* models predict a universe age between $2/3 T_0$ and T_0. For an apparent age T_0 of 15 billion years, this gives a true universe age of 10 to 15 billion years.

Astronomers are able to estimate the ages of stars. As shown in *Scientific American*, March 2001, (page 53), recent studies have found the ages of the oldest stars to be no less than 11.5 billion years and no greater than 14.5 billion years.

When we compare these stellar ages with a theoretical universe age between 10 and 15 billion years, it is obvious that the Big Bang theory has a serious age problem. Many versions of the Big Bang theory have been proposed to achieve a universe that is older than the oldest stars. However, even if the Big Bang theorists are sufficiently clever to surmount this hurdle, it is clear that the universe must have developed in an extremely efficient manner to have achieved its present state in the short time allowed since the Big Bang.

Measurement of Galaxy Distance

The Hubble constant is the ratio of galaxy velocity divided by galaxy distance. The galaxy velocity can be determined accurately from the galaxy spectrum, but the measurement of galaxy distance is extremely difficult. Hubble underestimated galaxy distances by a factor of 8, and so his estimate of the Hubble constant was 8 times too large. Let us see how Hubble determined galaxy distance.

Measurement of Parallax

The first step in measuring astronomical distances is to use parallax to measure the distances to nearby stars. Parallax is the principle that our eyes and brain use to achieve depth perception. Out two eyes receive different images. The brain compares the two images, and thereby distinguishes close objects from those that are further away.

You can observe the parallax effect by holding a finger a few inches in front of your nose, and looking with one eye at a time. As you switch between right and left eyes, the image of the finger moves back and forth relative to the distant background. The closer the finger is held to the eyes, the greater is the shift of the finger image relative to the background.

As the earth rotates around the sun, the parallax effect causes the images of nearby stars to shift relative to the distant stars. The parallax shift is inversely proportional to distance, and is equivalent to ±1.0 arc second for a distance of 3.26 light years. If the star image shifts by ±0.1 arc second over the year, we know that the star is located 32.6 light years away. By this means, astronomers can measure stellar distances out to about 100 light years with ground-based telescopes.

A star that is 100 light years away has a parallax shift of only ±0.03 arc seconds. Since this is a very small shift, the accuracy with which one can measure the distance of a star 100 light years away is poor.

The Hubble space telescope can make much more accurate parallax measurements, because it does not experience the blurring effect of the atmosphere. With the Hubble telescope, stellar distances have been measured more accurately from parallax, and out to greater distances.

There are about 100 stars within a distance of 20 light years, which indicates that the average distance between stars is about 7 light years. Since the volume of a sphere is proportional to the cube of its radius, the number of stars is approximately proportional to the cube of the distance. Consequently there are about 12,000 stars within a distance of 100 light years. This shows that astronomers can measure parallax to determine the approximate distances to about 12,000 stars.

The Cepheid Variable Stars

Astronomers have classified stars into different types according to their spectra. In many cases they can estimate from the spectrum of a star the approximate light power that it radiates, and this gives its approximate *absolute magnitude*. By comparing this with the observed magnitude of the star, they have estimated the distances of many stars much more distant than 100 light years.

An accurate yardstick for astronomy was developed by astronomer Henrietta Leavitt (1868-1921) in 1908, based on Cepheid variable stars. These are stars that vary periodically in luminosity, and are named after the star Delta Cephei. Leavitt observed the Cepheid variable stars located in the Small Magellanic Cloud, which is a cluster of stars outside our Milky Way galaxy, about 300,000 light years from the earth. Since all of the stars in the Small Magellanic Cloud are at approximately the same distance, they can be compared directly with one another.

Leavitt found that the period of variation of a Cepheid variable star is related to the light power that it radiates. By examining nearby Cepheid variable stars, the distance to which can be measured by parallax, she was able to relate the period of variation of a Cepheid variable star to the absolute light power that the star radiates. This was the key that allowed Edwin Hubble to make his revolutionary discovery of the expansion of the universe.

How Edwin Hubble Measured the Galaxy Distances

Hubble examined Cepheid variable stars in the galaxy M31 of the constellation Andromeda, which is 2.3 million light years away, and in the M33 galaxy of the constellation Triangulum, which is 2.6 million light years away. By this means he estimated the distances to M31 and M33. About 15 years later, astronomers discovered that there are two types of variable stars: the Cepheid variable stars and the RR Lyrae stars. This effect confused Hubble's measurement, and so he estimated the distances to M31 and M33 to be half the value that is accepted today.

Hubble examined the brightest stars in the M31 and M33 galaxies, and assumed these to be super-giant stars. He assumed that the brightest stars in more distant galaxies would radiate the same power as those in the M31 and M33 galaxies. With this approach he estimated distances to several galaxies more distant than M31 and M33, which were too far away to observe Cepheid variables.

Hubble categorized these galaxies into different types depending on their shapes. He assumed that all galaxies of a given shape have approximately the same dimensions. With this principle, he estimated the distances to galaxies that were much further away. He derived his universe expansion rate by comparing the redshifts with the estimated distances of these very distant galaxies

The universe expansion rate that Hubble measured was about 8 times greater than the value generally accepted today. As stated above, one serious problem was the confusion between Cepheid variable stars and RR Lyrae stars, which caused a factor of two error in the distance measurements. Another problem was that some of the "super-giant stars" that Hubble examined were actually regions of brightly illuminated hydrogen gas. There were many other sources of error that contributed to an inaccuracy factor of 8 in the initial Hubble constant.

Modern Measurements of the Hubble Constant

Measurements of the Hubble constant have improved greatly in recent years, assisted in particular with data from the Hubble Space Telescope. Two teams are working on the problem; one led by Alan Sandage and the other by Wendy Freedman.

The Freedman team has based its measurements on Cepheid variable stars. With modern telescopes, Cepheid variable stars can be observed in very distant galaxies. Sandage has based his measurements on supernovas. There is a particular type of supernova, called Type 1a, which emits nearly a fixed amount of power. Supernovas can be observed at very great distances.

The recent average Hubble constants reported by Sandage and Freedman are 18.7 and 21.5 km/sec per million light years [29]. Therefore a Hubble constant of 20 km/sec per million light years appears to be a good average value for the recent studies. On the other hand, there is appreciable uncertainty in these measurements, and the true value might lie significantly outside the range of these values

The Big Bang Concept

Material Contained in our Universe

The Big Bang theory postulates that our universe began as an extremely dense mass that exploded with a "Big Bang" about 15 billion years ago. The initial size of the universe predicted by modern Big Bang theorists is unbelievably small. To understand the implications of this miniscule initial size, let us calculate the amount of material in the present observable universe (as defined by the Big Bang theory) to determine how much matter was presumably jammed into the initial volume of the universe.

The size of the *observable universe* depends on the value of the Hubble constant H_0. As shown in item (1) of Table 4-1, the latest data gives an average Hubble constant H_0 of about 20 km/sec per million light years. Light year is denoted Lyr, and million light years is denoted MLyr. Based on this value for H_0, we have calculated the *apparent age* T_0 of the universe to be 15 billion years, as shown in item (2).

As defined by the Big Bang theory, the radius of the observable universe r_0 is equal to the distance that light can travel during the apparent age of the universe T_0. Since the apparent age is 15 billion years, the radius of the observable universe is 15 billion light years, as shown in item (3). If the whole universe expands at the rate of the Hubble constant, the radius r_0 of the observable universe is the distance at which a galaxy would travel at the speed of light.

Table 4-1: *Calculation of radius of Big Bang observable universe*

(1)	Hubble constant (H_0)	20 km/sec per MLyr
(2)	Apparent age of the universe (T_0)	15 billion years
(3)	Radius of observable universe (r_0)	15 billion light years
(4)	Light year (Lyr)	9.46×10^{12} km
(5)	Radius of observable universe (r_0)	142×10^{21} km

Item (4) in Table 4-1 gives the distance in kilometers that light travels in one year, which is denoted Lyr. (Note that 10^{12} means 1 followed by 12 zeros, which is one trillion. One trillion is 1000 billion and one billion is 1000 million.). Item (4) tells us that light travels 9.46 trillion kilometers in one year. Multiplying items (3), (4) gives the radius of the observable universe in kilometers, which is shown in item (5). In words, item (5) shows that the radius of the observable universe is 142 billion times one trillion kilometers (km).

The volume of the observable universe is equal to $(4/3)\pi r_0^3$, which is the well-known formula for the volume of a sphere. Applying this formula to the value for r_0 in item (3) of Table 4-1 gives the volume for the observable universe in item (1) of Table 4-2. The expression $MLyr^3$ represents a cube that is one million light years (MLyr) on a side.

Table 4-2: *Calculation of number of galaxies, number of equivalent suns, and total matter of Big Bang observable universe*

(1)	Volume of Observable Universe	14.1×10^{12} $MLyr^3$
(2)	Average distance between galaxies	10 MLyr
(3)	Average volume per galaxy	1000 $MLyr^3$
(4)	Number of galaxies	14×10^9 (14 billion)
(5)	Luminosity density in suns	5.1×10^6 per $MLyr^3$
(6)	Average suns per galaxy	5.1×10^9 (5.1 billion)
(7)	Suns in observable universe	71×10^{18} (71 billion, billion)
(8)	(total mass)/(luminous mass)	100
(9)	Number of sun masses in universe	7.1×10^{21} (7.1 billion trillion)

Silk [21] (p. 396) reports that the typical distance between galaxies is 10 million light years (10 MLyr), as shown in item (2) of Table 4-2. Hence we can allot to an average galaxy a cubic volume that is 10 million light years on a side. As shown in item (3) the volume of this average cube is 1000 $MLyr^3$. The volume of the universe in item (1) is divided by the volume per galaxy in (3) to obtain the estimated number of galaxies in the universe shown in item (4). Since 10^9 represents one billion, item (4) shows that the observable universe is estimated to contain 14 billion galaxies.

Item (5) in Table 4-2 shows that the average radiation within a volume of one $MLyr^3$ is equal to that from 5.1×10^6 (5.1 million) suns. This was obtained from Table 11-1 (item 6) of *Believe* [1], and was derived from astronomical data in the *Revised Shapley-Ames Catalogue*. Multiplying item (5) by the average volume per galaxy in item (3) gives the average luminosity per galaxy shown in item (6). This shows that the average radiation per galaxy is equal to that from 5.1 billion suns. For comparison, our Milky Way galaxy contains luminous material equivalent to 10 billion suns. Since our Milky Way galaxy is a large galaxy, the value in item (6) for an average galaxy appears to be reasonable.

The number of galaxies in item (4) is multiplied by the average suns per galaxy in item (6) to obtain the equivalent radiation of the whole universe in item (7), which is expressed in terms of equivalent suns. Item (7) shows that the light radiated from the observable universe is equal to the radiation from 71×10^{18} suns. In words, this can be expressed as 71 billion times the radiation from one billion suns. Since radiation is approximately proportional to mass, this says that the mass of the stars that are shining in our observable universe is 71×10^{18} times the mass of our sun.

There is much more dark matter in our universe (which we cannot see) as there is luminous matter (which we can see). Studies of the rotation of our own Milky Way galaxy, and the rotation of other galaxies have shown that there must be about 10 times as much dark matter in a galaxy as there is luminous matter, in order for the different parts of the galaxy to rotate at their observed velocities. Studies of clusters of galaxies have shown that there should be about 400 times as much dark matter associated with the cluster as there is luminous matter.

This dark matter might consist of diffuse gas in space, black dwarf and neutron stars, planets, and brown dwarf stars that are too small to ignite.

We assume that there is about 100 times as much dark matter in space as there is luminous matter, which is a conservative estimate. Item (8) of Table 4-2 shows that we estimate the total mass of the universe to be 100 times the luminous mass. Hence item (7) is multiplied by 100 to obtain the total mass of the observable universe, shown in item (9), which is 7.1×10^{21} times the mass of our sun.

These calculations indicate that our observable universe has about 14 billion galaxies, which contain luminous material equivalent to 71 billion times one billion suns. The total mass of the universe is estimated to be 7.1 billion times the mass of one trillion suns.

The Neutron Star Model of Universe Creation

Let us assume that at the instant of the Big Bang all of the material of our universe was compressed to the density of a neutron star. At this density the material would have been compressed as tightly as is allowed by our normal laws of physics. The matter of our universe would have consisted entirely of tightly packed neutrons. The universe would have been one gigantic neutron star.

Before determining the size of this neutron-star universe, it is convenient to consider an intermediate model. Let us assume that the

4. The Creation of Our Universe 69

universe is squeezed into a single sphere having the density of water. The size of this water-density universe is calculated in Table 4-3. Item (1) shows the total mass of the universe in terms of equivalent suns, which was obtained from item (9) of Table 4-2.

Table 4-3. *Calculation of radius of neutron-star model of initial universe*

(1) Total mass of universe	7.1×10^{21} suns
(2) Radius of sun-like star with water density	780,000 km
(3) Radius of universe with water density	15.0×10^{12} km = 1.58 Lyr
Approximate	14.2×10^{12} km = 1.5 Lyr
(4) Radial compression, water to neutron star	65,000
(5) Radius of neutron-star universe	218 million km
(6) Mean radius of Earth orbit	150 million km
(7) Mean radius of Mars orbit	228 million km

Our sun has a radius of 700,000 km and an average density that is 1.4 times that of water. Multiplying the sun radius by the cube root of 1.4 gives 780,000 km. As shown in item (2) of Table 4-3, this is the radius of a star with the density of water that has the mass of our sun. If we multiply item (2) by the cube root of item (1), we have in item (3) the radius of our whole universe if it were squeezed to form a single body with the density of water. This is a theoretical model, which ignores the effect of gravitational forces on such a large body.

The radius in item (3) is 15.0×10^{12} km. Since one light year is 9.47×10^{12} km, this radius is equal to 1.58 light years. To simplify our discussion, it is convenient to round off this radius to 1.5 light years.

Mould [7] (p. 350) specifies the density of a neutron star to be 2×10^{14} times the average density of the sun. Mould calls this the density of "nuclear matter". This represents a density of 2.8×10^{14} times that of water. Gamow [14] (p. 332) gives the "density of nuclear fluid" as 10^{14} times that of water. However, this appears to be merely a rough approximation, and so we assume the value given by Mould.

The cube root of 2.8×10^{14} is 65,420. This shows that the diameter of a star is reduced by a factor of 65,420 if it is compressed from the density of water to that of a neutron star. To simplify our discussion, we round off this factor to 65,000, which is shown in item (4) of Table 4-3.

Dividing the approximate value of item (3) by item (4) shows in item (5) that the observable universe would have a radius of 218 million km, if it were compressed to the density of a neutron star.

Items (6) and (7) show for reference the mean radii of the orbits of Earth and Mars. Comparing these with item (5) shows that the neutron-star universe would have a radius considerably larger that that of the orbit of Earth, and nearly equal to that of the orbit of Mars.

We can simplify these results with the approximate summary given in Table 4-4. Item (1) shows that the observable universe has a diameter of 30 billion light years. We assume that the enormous space between molecules in the universe is eliminated to form a single body with the density of water. The observable universe would be squeezed to a diameter of 3 light years, as shown in item (2).

Table 4-4. Summary of calculation of size of neutron-star model of initial universe

(1) Diameter of observable universe   30 billion light years
(2) Diameter of universe with water density   3 light years
(3) Radial compression, water to neutron star   65,000
(4) Diameter of neutron star with universe mass   436 million km
(which approximates the diameter of the Mars orbit)

Now assume that the universe is squeezed so that it consists entirely of tightly packed neutrons. This step would eliminate the space inside the atom. As shown in item (4), the resultant neutron-star universe would have approximately the diameter of the orbit of the planet Mars. If one were to apply the Big Bang theory, this would be a realistic physical assumption of the size of the observable universe at the instant of the Big Bang.

George Gamow, the Father of the Big Bang

In 1917, the year after Einstein presented his General theory of Relativity, cosmological models based on the Einstein theory were independently published by Einstein and Willem de Sitter (1872-1934), a Dutch astronomer. These were static cosmological models that did not change with time. Einstein included in his gravitational field equation an arbitrary "cosmological term", which allowed the equation to yield a static cosmological solution.

When Hubble discovered the expansion of the universe in 1929, Einstein realized that these static cosmological models could not explain the Hubble expansion. A dynamic solution of his gravitational field equation was needed, and a dynamic solution would not require the arbitrary cosmological term. At that point Einstein disavowed his cosmological term, claiming it to be, *"one of the greatest mistakes of my life"*.

Time-varying cosmological models based on the Einstein theory were developed independently by Alexander Friedmann (1888-1925) in 1922, and by Georges Lemaitre (1894-1966) in 1927. Friedmann was a Russian meteorologist and Lemaitre was a Belgian cleric of the Roman Catholic Church. Friedmann died in 1925 after being chilled in a weather balloon.

In the 1930's Lemaitre applied his cosmological model to explain the creation of the universe. This model implied that the universe was created as an extremely dense mass, which exploded and has been expanding ever since. However, Lemaitre faced the serious problem that the accepted value of the Hubble constant at that time was 8 times too large, and so the apparent age of his universe model was only 2 billion years.

During World War II little attention was paid to cosmological models. When the war ended, George Gamow (1904-1968), who had worked on the Manhattan atomic bomb project, became a strong proponent of the Lemaitre concept. He applied his knowledge of physics to astronomy and wrote books for the general public that became very popular, such as Refs. [14, 15]. He widely publicized the notion that our whole universe began in an enormous explosion billions of years ago. At that time the accepted value for the Hubble constant had been greatly reduced, and the apparent age of the universe had increased to 5 billion years.

An alternative to the Lemaitre-Gamow cosmology theory was proposed in 1948 by astrophysicists Fred Hoyle, Hermann Bondi, and Thomas Gold, which was called the Steady-State Universe theory. This theory assumes that the universe is infinitely old, and that diffuse matter is being created throughout the universe to compensate for the Hubble expansion of the universe.

As a criticism of the Lemaitre-Gamow cosmological theory, Hoyle used the term "Big Bang" to characterize the enormous explosion that presumably created our universe. The name stuck, and since that time, the Lemaitre-Gamow cosmology model became known as the "Big Bang theory".

Gamow proposed that the density of the universe at the time of the Big Bang was the same as that of a neutron star. He assumed that the universe began as an enormous neutron star. In his book, *One, Two, Three, Infinity* [14] (page 337), he calculated the initial size that would have expanded to a radius of 500 million light years. That was the range of the Mount Wilson Observatory, and Gamow knew that galaxies extended to that distance. This initial size had a radius of 8 times the sun radius, or 5.6 million kilometers.

At the time that Gamow wrote his book, the Hubble constant implied an apparent universe age of 5 billion years, so that the radius of the observable universe was 5 billion light years. (See Ref. [14], page 337.) If galaxies extend throughout the observable universe, Gamow's limit should be increased by a factor of 10 from 500 million light years (the range of the Mount Wilson telescope) to 5 billion light years (the radius of the observable universe). Therefore, Gamow's estimate of the initial universe at the time of the Big Bang would have had a radius of 80 sun radii, or 56 million kilometers, if it had included the total observable universe.

This estimate of 56 million kilometers is somewhat smaller than our 218 million kilometer estimate, but the numbers are consistent. Our estimate included a 100-to-1 ratio of dark matter to luminous matter, which Gamow did not consider.

The size of the initial universe at the time of the Big Bang that was shown in Table 4-4 is consistent with the estimate that Gamow gave, if modern astronomical data are applied. **The Gamow Big Bang concept assumed that the universe had the density of a neutron star at the instant of the Big Bang.**

The Singularity Model of Universe Creation

The Gamow concept was the accepted Big Bang theory until the mid 1960's. At that time, powerful computers became widely available. Many scientists found that computer analysis of the Einstein General theory of Relativity was a rewarding endeavor to advance their careers. Since the aura of the Einstein name is enormous, scientists who engaged in this research were greatly rewarded.

With computers, the complicated Einstein gravitational field equation could now be solved in a manner unheard of in Einstein's day. About the only area where these studies could be applied was cosmology. Hence an enormous effort in cosmological research began, and this research led almost invariably to the Big Bang theory. Since 1980 this research has involved many hundreds of scientists.

This new Big Bang research, which was based on computer studies of the Einstein gravitational field equation, resulted in the "singularity" concept. This concept was expressed as follows in a book by Filkin [19]

4. The Creation of Our Universe 73

(p. 104), titled *Stephen Hawking's Universe*:

> *"Stephen [Hawking] and Roger Penrose published a paper in 1970 which proved that, if Einstein's mathematics were correct, a singularity had to result from a black hole, and had to exist at the start of the universe. - - - The paper argued that if relativity as explained by Einstein is correct — and all of the evidence from observation seems to keep confirming it — then the universe must have started with a big bang explosion out of a singularity. The equations do not allow an alternative."*

Filkin directed a 1997 television series for the Public Broadcasting System, also titled, *Stephen Hawking's Universe*. This book was a supplement to that television series.

What is meant by a "singularity"? Joseph Silk is the author of three books on the Big Bang theory. One of these [22] was officially endorsed by the *Scientific American*, and is a volume in the *Scientific American Library*. This book [22] gives on page 66 the following definition of a singularity, as recognized by Big Bang cosmologists:

> ***Singularities:*** *The universe began in time zero in a state of infinite density. At least the existence of such a singular state is the expectation from extrapolating the present universe backward in time. Of course, the phrase 'a state of infinite density' is completely unacceptable as a physical description of the universe. An infinitely dense universe would be what is called a 'singularity', where the laws of physics, and even space and time, break down. The resolution of the paradox is that our theory of gravitation has broken down before reaching this extreme state. The first 10^{-43} sec is inaccessible in our current theories. The end of the period, called the Planck instant, represents the beginning of time according to those theories. Nevertheless, to the extent that one accepts Einstein's theory of gravity, singularities are predicted to exist in nature."*

Silk [22] (p. 79) illustrates two versions of the modern Big Bang theory, which start 10^{-35} sec after the Big Bang. One of these applies the "inflation" postulate.

Silk explains that recent Big Bang models have favored the "inflation" concept, which was first proposed in a preliminary form by Alan Guth in 1981. This concept postulates that in the initial period the universe size doubled every 10^{-35} second, until it reached a much larger size. Then the rate of expansion slowed.

Silk [22] illustrates on page 79 the early (pre-inflationary) Big Bang concept and a concept that applies the inflation postulate. Both of these started 10^{-35} sec after the Big Bang. In the early Big Bang model, the observable universe began as a body that was 1 mm in diameter. In the inflationary Big Bang model, the observable universe began with an initial diameter of 10^{-24} mm.

The initial size of the model with inflation is so incredible that we disregard it. This leaves us with a diameter of 1 millimeter as the initial size of the observable universe given by Silk.

Another recognized authority on the Big Bang theory is P. James E. Peebles. As stated earlier, the Jan. 2001 *Scientific American* (page 37) called Peebles the *"father of modern cosmology"*. The October 1994 *Scientific American* (p. 53) gave an article by Peebles and others, which began with:

> *"At a particular instant, roughly 15 billion years ago, all of the matter and energy we can observe, concentrated in a region smaller than a dime, began to expand and cool at an incredibly rapid rate."*

Since a dime is 18 mm in diameter, Peebles is claiming that the initial size of the observable universe was less than 18 mm.

Of our three experts on the modern Big Bang theory (Hawking, Silk, and Peebles), Peebles has given the largest estimate for the initial size of the observable universe. Peebles claimed that the initial diameter of the observable universe was less than 18 mm.

Now let us return to the initial Big Bang universe proposed by Gamow, the "father of the Big Bang theory". ***At the time of the Big Bang, Gamow postulated that the universe had the density of a neutron star. To have this density, the observable universe would have been the size of the orbit of Mars.***

If our present universe were compressed to remove the enormous spaces between molecules, our observable universe would be reduced to a body with the density of water and a diameter of 3 light years. If the diameter of this body were compressed further by a factor of 65,000, the body would have the neutron-star density postulated by Gamow at the instant of the Big Bang. It would have the diameter of the orbit of Mars.

Modern Big Bang cosmologists predict a very much smaller size of the universe at the instant of the Big Bang. They claim that matter can collapse to a density that is very much greater than that of a neutron star.

For the purpose of discussion, let us assume that there is an unknown physical process that allows a body of neutron star density to collapse to a very much greater density. Let us make the extreme physical postulate that a body of neutron-star density can be compressed in radius by a factor of 65,000. Relative to neutron-star density, the density of this body would increase by the same factor as a body that is compressed from water density to neutron-star density.

With this extreme postulate, the diameter of our neutron-star universe (436 million km) would be compressed by 65,000 to a diameter of 6700 km, which is approximately the diameter of the planet Mars (6800 km). This initial universe would still be enormous in comparison to the size predicted by modern Big Bang cosmologists. To reduce this to the 18-mm size of *"a dime"* predicted by Peebles would require another factor of 370 million in radial compression.

Let us summarize these results. If we apply the Gamow hypothesis that the universe had the density of a neutron star at the instant of the Big Bang, our observable universe would have been the size of the orbit of Mars. We then make the extreme (and unwarranted) postulate that this neutron-star universe can be compressed by the same factor that water is compressed to achieve the density of a neutron star. This would have produced a body the size of the planet Mars. Even with this extreme postulate, the diameter of the initial universe would still be hundreds of millions of times larger than the diameter predicted by modern Big Bang cosmologists.

This shows that modern Big Bang cosmologists are not dealing with physical reality. They are operating in fantasy-land.

Modern Big Bang cosmologists are deriving their conclusions from the Einstein gravitational field equation, which they consider to represent absolute truth. They are finding singularity effects in the mathematics, and are insisting that these mathematical singularities must occur in physical reality. They are using the enormous prestige of Albert Einstein to support their physical predictions, but are absolutely rejecting the philosophy that Einstein demanded throughout his career.

As shown in Chapter 3 (p. 54), Einstein drastically opposed the black-hole singularity (which he called the "Schwartzschild singularity"). During Einstein's lifetime, the Big Bang theory used the Gamow concept, which assumed that the initial universe had the density of a neutron star. ***Neither the black-hole singularity nor the Big Bang singularity was claimed by responsible scientists while Einstein was alive.***

Modern Big Bang cosmologists say that they are following in the footsteps of Einstein, because they have derived their black hole and

Big-Bang singularity concepts from the Einstein gravitational field equation. Although Einstein strongly rejected the black hole singularity, they assert that they have proven with computer studies that the Einstein theory does indeed predict such a singularity.

Nevertheless, in 1945 Einstein had considered the concept that our universe may have begun as a singularity at the instant of the "Big Bang". He flatly rejected this interpretation of his theory with the following [66] (p. 129):

> *"Theoretical doubts [concerning the creation of the universe] are based on the fact that [at the] beginning of the expansion, the metric becomes singular and the density becomes infinite. . . In reality, space will probably be of a uniform character, and the present [relativity] theory will be valid only as a limiting case. . . One may not therefore assume the validity of the equations for very high density of field and of matter, and one may not conclude that the 'beginning of the expansion' must mean a singularity in the mathematical sense. All we have to realize is that the equations may not be continued over such regions."*

The full wording of this statement is given at the end of Appendix E.

This 1945 quotation shows that Einstein strongly opposed the "Big Bang" singularity, and the 1939 quotation [13] given in Chapter 3 (p. 54) shows that Einstein strongly opposed the "black hole" singularity. *In summary, Einstein definitely rejected the "black hole" singularity in 1939, and he definitely rejected the "Big Bang" singularity in 1945.* Even if Einstein had seen computer evidence that his theory predicts a singularity, there is no reason to believe that this would have altered his firm conviction that *"singularities do not exist in physical reality"*.

Cosmic Microwave Background Radiation

An important milestone in the development of the Big Bang theory occurred in 1965, when Arno Penzias and Robert Wilson, two physicists working at Bell Laboratories, discovered cosmic microwave background radiation. These physicists were performing measurements on a sensitive microwave antenna that had been developed for use in satellite communication. When communication satellites generated greater power, there was no longer a communication need for this sensitive instrument, and so the antenna was redirected toward basic research.

Penzias and Wilson discovered spurious signals in their antenna at microwave frequencies, which they could not explain. These signals were due to cosmic radiation coming from all directions in space. Big Bang theorists had predicted that such radiation should have been generated by the Big Bang.

Gamow had predicted that optical radiation from the early universe should still be observable today. With the expansion of the universe, this cosmic optical radiation should be reduced in frequency, and should be observable today at microwave frequencies. He predicted cosmic radiation coming from all directions that is equivalent to the radiation from an ideal blackbody. Gamow made several estimates of the effective temperature of this cosmic blackbody radiation, which varied from 5 degrees Kelvin to 20 degrees Kelvin.

The radiation from an ideal blackbody was discussed in Chapter 3. General plots for the spectrum of this radiation were given in Fig. 3-6 on page 57. An equation associated with Fig. 3-6 shows that the intensity of radiation varies as the fourth power of the absolute temperature. Consequently, the 4-to-one temperature range from 5 to 20 degrees Kelvin predicted by Gamow corresponds to an intensity ratio of 4^4, which is 256. Thus the Gamow prediction of blackbody radiation was not very precise. There was a factor of 256 uncertainty in the predicted intensity of the blackbody radiation.

In the early 1960's, Prof. Robert Dicke of Princeton University was studying the Big Bang theory. One of his graduate students, P. James E. Peebles, investigated Gamow's prediction of cosmic microwave radiation, and considered building an antenna to measure this radiation. Peebles estimated that the blackbody temperature should be 30 degrees Kelvin. Then he discovered that Penzias and Wilson had already measured similar radiation in their antenna. The antenna had detected unexplained electrical disturbance signals with a spectrum that corresponded to a blackbody temperature of 3.5 degrees Kelvin.

This discovery of the cosmic microwave background radiation was loudly publicized by Big Bang theorists. Since this cosmic radiation had been predicted by the Big Bang theory, it was proclaimed to be proof that the Big Bang theory was correct. However, the Big Bang proponents have failed to mention that this radiation was only predicted in a qualitative sense. Since radiation intensity varies as the fourth power of temperature, the 30 degree Kelvin estimate by Peebles corresponds to a radiation intensity that is 5000 times greater than the intensity for a blackbody at 3.5 degrees Kelvin.

In 1948, Ralph Alpher and Robert Herman, two graduate students working with George Gamow, had published a paper in the scientific journal *Nature* that predicted cosmic microwave radiation corresponding to a blackbody temperature of 5 degrees Kelvin. Since this is the closest estimate to the measured temperature, it is the value generally quoted by Big Bang proponents.

Much more accurate measurements of this cosmic radiation were obtained from the Cosmic Background Explorer (COBE) satellite in 1989. This satellite measured cosmic microwave radiation coming uniformly from all directions that corresponded very accurately, in intensity as well as in spectrum, to the radiation from an ideal blackbody at a temperature of 2.73 degrees Kelvin.

The Big Bang theory claims that this cosmic microwave background radiation is the cooled relic of optical radiation that was emitted from the early universe about 300,000 years after the Big Bang. However, an important weakness of this assumption is that the cosmic radiation emanates with extreme uniformity from all directions. Lerner [18] (p. 31) reports that this cosmic radiation is far too uniform to explain the lumpy arrangement of galaxies in the universe that has been observed in recent universe surveys. In the data obtained from the COBE satellite, the energy of radiation received from different directions varies by only a few parts in 100,000.

This implies extreme uniformity of the universe at this early period. How did this highly uniform early universe create what we observe today? Not only is our universe separated into galaxies, but the galaxies are not spaced uniformly. As we will see, the galaxies are arranged into long curling filaments, which form huge ribbon-like structures that are typically 100 million light years thick and 300 million light years wide.

At the time that cosmic microwave background radiation was discovered by Penzias and Wilson in 1965, both the Steady-State Universe theory and the Big Bang theory were seriously considered by scientists as possible explanations for the Hubble expansion. The Steady-State Universe theory had been proposed in 1948 by Fred Hoyle, Hermann Bondi, and Thomas Gold. This theory did not have an immediate explanation for this cosmic radiation, and so interest in the Steady-State Universe theory declined rapidly.

This event is recounted in the recent book by Hoyle, Burbidge and Narlikar [20] (pp. 84-85). This book concludes in the following (p. 84) that the primary impetus toward the Big Bang theory was economic

> *"[The 1965 discovery by Penzias and Wilson] coincided with the beginning of a great expansion of the number working in astronomy, an expansion begun by the intense interest at the time in space,*

4. The Creation of Our Universe   79

which led to NASA being funded in billions of US dollars, rather than in the few tens of millions that had been previously available to astronomy, even in the most favorable circumstances. The number of attendees at the International Astronomical Union jumped in only a few years from 300 or 400 to more than 2000, an immense expansion that made it possible to claim that cosmology as a branch of physical science only began in earnest in 1965."

Thus, Hoyle, et al, recognize that the stampede to the Big Bang, and the eclipse of the Steady-State Universe theory, was caused primarily by economic forces that happened to coincide with the discovery of the cosmic microwave background radiation by Penzias and Wilson.

As this book shows, another major factor was the wide availability of powerful computers in the mid 1960's. Many scientists were induced to perform computer studies of the Einstein gravitational field equation, and these studies led almost invariably to the Big Bang theory.

When support for the Steady-State Universe theory waned, Bondi and Gold lost interest, and with time Hoyle gradually abandoned his theory.

This book shows in Appendix C that a cosmology model based on the Yilmaz theory also predicts cosmic microwave background radiation. The predicted blackbody temperature lies in the range between 2.1 and 3.4 degrees Kelvin. This range is much closer to the 2.73 degree Kelvin temperature measured by the COBE satellite than the best estimate (5 degrees Kelvin) made by Big Bang theorists.

The Quasar

The Discovery of the Quasar

After World War II, a great many radio antenna systems were constructed to act as "radio telescopes" that detected the radio emissions from the heavens. These investigations led to the discovery of a strange new type of star, which was called a *quasi-stellar radio object*, and was given the acronym *quasar*.

The resolution of radio telescopes is poor because of the long wavelengths of the signals. Much higher resolution was achieved for quasars eclipsed by the moon. The instant at which a quasar disappeared behind the moon was detected. The radio signal was compared with a nearby optical star image eclipsed by the moon at the same instant. This allowed a radio signal to be related accurately to an optical star image.

Astronomers began to study the optical spectra of these radio objects, and found them to be peculiar. The spectra did not look like the spectra of known elements. Finally in Feb. 1963, astronomers Jesse L. Greenstein and Maarten Schmidt jointly discovered that they could explain the quasar spectra by assuming that quasars have extremely large spectral redshifts.

It was soon discovered that not all of these strange stars with extreme redshifts were radio sources. Consequently their name was changed from *quasi-stellar radio objects* to *quasi-stellar objects*, but the acronym *quasar* was retained. They are also called *QSO's*.

The anomalous redshift of the quasar is a serious enigma. If this redshift is a Doppler effect caused by velocity, the quasars must be receding from us at extremely high velocities. This in turn requires them to be billions of light years away. For quasars to be at such vast distances, they must radiate enormous power for us to see them.

For example, one of the first two quasars to be studied is quasar 3C48, which is examined in Appendix B. It has a redshift of 0.367, which corresponds to a Doppler velocity of approximately 110,000 km/sec. By assuming a Hubble constant of 30.7 km/sec per million light years, Greenstein and Schmidt estimated the distance to this quasar to be 3.6 billion light years. At that distance the quasar would have to radiate the power of 1000 billion suns to achieve the stellar magnitude that we observe This radiation is 100 times the total power radiated by our enormous Milky way Galaxy.

Then they discovered that the power of 3C48 varied by 40 percent over a period of 600 days. This rapid variation indicates that 40 percent of this enormous quasar power must be generated within a volume that is only a few light years thick. The quasar is indeed a strange object.

Later studies found quasars with much larger redshifts. Studies have also shown that some quasars vary in power over much shorter periods. Quasars have been discovered that vary by a factor of 2 in power within several hours, and so can be no larger than our solar system.

Quasar Observations of Halton Arp

Because of the unbelievable properties attributed to quasars, the noted astronomer, Halton Arp, suspected that they may be much closer than was being assumed. He started making observations of quasars to obtain direct astronomical estimates of their distances.

Arp found many quasars with images very close to galaxies having much smaller redshifts. If the image of one quasar is very close to that of a galaxy, this might be a chance relationship. The quasar might be billions of light years beyond the galaxy. However, if two or more quasars appear to be close to a galaxy, the probability of a chance relationship is remote. We ask, "What is the probability that the images of two or more quasars would fall this close to an arbitrary direction in space?" Probability considerations indicate that it is highly unlikely that the quasars are not physically close to the associated galaxy.

Arp photographed three quasars that appear to be in the outer fringe of galaxy NGC 3842. It is theoretically possible that this is a chance relationship; that the quasars actually lay far beyond the galaxy. However, the probability that three quasar images would fall this close to an arbitrary direction in space is about one in a million.

Arp found several cases like this. The possibility is essentially zero that all of these observations could be accidental, which relate quasars to galaxies having much smaller redshifts.

In Halton Arp's study of quasars, he also found many cases of filament structures that directly connect quasars to galaxies having much smaller redshifts. These observations appear to show that the quasar was ejected from the associated galaxy by a supernova explosion. There are several cases of a filament connecting a quasar to a galaxy, and another opposing filament on the other side of the galaxy. This suggests that the opposing filament is the reaction from a supernova explosion that ejected the quasar from the galaxy.

The quasar observations by Arp were opposed by the astronomical community, because they did not agree with the accepted dogma that quasars are billions of light years away. He found it very difficult to get his findings published. Some journals rejected his work, and papers were often held up for years by referees. Finally in 1984, the committee that controls observation time at Palomar and Mount Wilson Observatories refused to allow Arp to use these facilities.

Dr. Halton Arp had performed distinguished research at Palomar and Mount Wilson Observatories since he received his PhD degree in 1953. He was president of the Astronomical Society of the Pacific from 1980 to 1983, and received awards from the American Astronomical Society, the American Association for the Advancement of Science, and the Alexander von Humbolt Senior Scientist Award. After being denied research facilities in California, he was forced to move to Germany to continue his career, where he joined the Max Planck Institute for Physics and Astrophysics in Munich.

Arp presented his quasar observations up to the time of his expulsion in his 1987 book, *Quasars, Redshifts, and Controversies* [16]. In his later 1998 book, *Seeing Red* [17], he also includes the extensive observations on quasars that he has made since he was forced to move to Germany. Arp's latest book gives overwhelming evidence that quasars are very much closer than is generally assumed. Nevertheless, officials in astronomy completely ignore this evidence, and continue to insist that the conventional quasar dogma is correct. A detailed discussion of the quasar is given in Chapter 11.

Implications of Gravitational Theory

The Big Bang theory of George Gamow makes a physically realistic postulate. It assumes that our universe began as a compact body with the density of a neutron star. At the instant of the Big Bang, our observable universe would have been compressed into a body with neutron-star density having the size of the orbit of Mars.

You may ask, "Why have modern Big Bang cosmologists abandoned the realistic Gamow concept to invoke their physically impossible singularity concept? Why do modern cosmologists insist that the initial universe was so small as to be devoid of physical reality?"

The reason is that modern Big Bang research is based on studies of the Einstein General theory of Relativity. Computer analyses of the Einstein gravitational field equation have been applied to cosmology, and these analyses have led to the Big Bang singularity concept.

This tells us that in order to investigate the creation of our universe, we must understand the Einstein General theory of Relativity.

"But how can I do that?" you may ask. "After all, it is well recognized that the Einstein theory is far beyond the comprehension of anyone who is not a highly trained scientist."

This book refutes that myth. Even though the mathematical calculations of General Relativity are very complicated, the physical principles underlying those calculations can be presented in a simple manner. This book will explain Einstein's Special and General theories of Relativity in terms that can be readily comprehended by the average reader. This should remove the mystique that surrounds the Einstein theory and give the reader the insight to realize (in agreement with Einstein) that, *"singularities do not exist in physical reality"*.

The Einstein General theory of Relativity is basically a theory of gravity, and so should be properly called the Einstein theory of gravity. As a preliminary step in investigating the Einstein theory we must first understand the gravitational theory of Isaac Newton.

Chapter 5

Newton's Theory of Gravity

Development of Newton's Theory

The theory of gravity developed by Isaac Newton (1642-1727) was based on the findings of Galileo and Kepler. Galileo had studied the characteristics of falling bodies, and discovered that a body falls to earth with a constant acceleration, which we call the *acceleration of gravity*. This was a key principle in Newton's theory. Newton applied his theory to the motions of planets in our solar system, and showed that his theory exactly satisfies the empirical laws of planetary motion that had been developed by Kepler.

The foundation for this theory was Newton's invention of calculus. Newton applied calculus to the findings of Kepler and Galileo and thereby developed the following two laws of mechanics:

*(1) **Law of gravitational attraction:** Two bodies are attracted together with a force that is proportional to the product of their masses, and is inversely proportional to the square of the distance between their centers of gravity.*

*(2) **Law of motion:** The force applied to a body is equal to the mass of the body times its acceleration, where acceleration is the rate-of-change of velocity. An alternative statement of this is: the force applied to a body is equal to the rate-of-change of momentum, where momentum is the product of mass times velocity.*

Newton also gave the following laws, which are corollaries of his two basic laws:

(3) A body at rest, or moving at constant velocity, stays in that condition unless a force is applied to it.

(4) For every action there is an equal and opposite reaction.

By applying calculus to these laws, Newton was able to calculate accurately the motions of bodies in our solar system. Newton's theory was published in 1687 by the Royal Society of England as *Philosophiae Naturalis Principia Mathematica*. Latin was used because scientists in all countries could read Latin. The famous scientist, Edmund Halley (1657-1742), personally paid for the printing. An English translation of Newton's *Principia* is given in Reference [62].

Newton's invention of calculus was the key that allowed Newton to develop his physical laws. Nevertheless, he did not use calculus in his *Principia*, because other scientists did not understand it. Instead, Newton applied graphical constructions that achieved the effects of calculus.

It is commonly believed that Newton discovered the principle of gravitational attraction, but this is not true. Forty-two years before Newton published his *Principia*, Ismaelis Bouillard had postulated that mutual attraction of the planets varies inversely as the square of the distance between them. [59] It is probable that even before that, both Galileo and Kepler understood that gravitational attraction holds the planets in their orbits around the sun, and holds the moon in its orbit around the earth.

What Newton actually developed was a precise set of mathematical laws, with the proof that these laws accurately describe the orbits of bodies in our solar system.

The Heliocentric Theory of Copernicus

In the 1600's, Galileo, Kepler, and Newton made great advances in the understanding of the motions of bodies in our solar system. Their achievements were based on the revolutionary *heliocentric* theory published by Nicholaus Copernicus in 1543. Nicholaus Copernicus (1473-1543) was a Polish astronomer, whose Polish name was Mikolai Kopernik. He worked as clergyman (possibly a priest), secretary, and physician at the Frauenburg cathedral in Polish East Prussia.

Since the time of the ancient Greeks, there had been strong belief in the *geocentric* (earth centered) concept of Aristotle and Ptolemy (c. 100-170 AD) that the sun and planets revolve around the earth. Copernicus presented astronomical data to support his *heliocentric* (sun-centered) theory that the earth and planets revolve around the sun.

The manuscript describing the Copernicus theory was completed by 1530, but was not published until 1543, just before Copernicus died. A preliminary version of this theory had received harsh criticism from many sources, including Protestant leaders Luther and Calvin, but Pope Leo X had expressed open-minded interest. Copernicus withheld publication of his full document because he was afraid of the severe reaction it would receive. He agreed to let friends publish it when he knew he was close to death. [57]

Galileo and his Telescope

About 1608 a spectacle maker in Holland built the first telescope. Galileo, a professor in Italy whose full name was Galileo Galilei (1564-1642), heard of this and began making his own telescopes. In 1610 Galileo revolutionized astronomy with his telescope. He studied our moon and found that it has mountains and valleys like the earth. He looked at Jupiter and made the fantastic discovery that Jupiter has moons of its own. He also saw that Venus has phases like our moon, which showed that Venus must revolve around the sun. Galileo published his findings, and claimed that his astronomical observations proved that the *heliocentric* theory of Copernicus must be correct.

Galileo received strong criticism for his support of the Copernicus concept from a number of influential university professors and clergymen. Galileo contributed to this conflict by writing material that was taken as a personal affront by politically powerful intellectuals. They finally convinced the Catholic Church to bring Galileo to trial. In 1633 Galileo was convicted of heresy and imprisoned.

However Galileo was soon placed under house arrest, and lived comfortably in his villa near Florence. His nun daughter lived with him, and he was able to receive visitors and teach students. He wrote defiant books that were smuggled out to foreign publishers. He died in 1642, the year that Isaac Newton was born. [58, 60]

The Galileo incident is often considered to be part of a continual conflict between science and Christianity, particularly the Roman Catholic Church, but this concept is simplistic. Galileo's book *Dialogue*, published in 1632, infuriated powerful intellectual leaders. These intellectuals used their political power to have Galileo's voice suppressed by the government. In Florence, Italy at the time of Galileo, the Catholic Church was the government.

Between 1450 and 1700, thousands of people were executed

throughout Europe for witchcraft, in both Protestant and Catholic areas. When compared with this, Galileo's punishment was not as harsh as it may seem today.

The Kepler Laws of Planetary Orbits

Another strong supporter of the Copernicus concept at the time of Galileo was Johannes Kepler (1571-1630), a German astronomer. In 1601 Kepler became director of Tycho Brahe's observatory near Prague, after Brahe died. For many years, Tycho Brahe (1546-1601), a Danish astronomer, had made accurate measurements of planet and star locations. With very large instruments having no optical magnification, Brahe made optical measurements to a precision of 10 arc seconds.

Kepler applied Brahe's data to the Copernicus model, and derived three accurate relations for the orbits of the planets, which are known as Kepler's Laws. These are:

(1) A planet orbits the sun in an elliptical orbit, with the sun at one focus of the ellipse.

(2) A planet moves more rapidly when nearer the sun than when further away, such that a radius drawn from the planet to the sun sweeps over an equal area for an equal time interval.

(3) The expression r^3/T^2 is the same for all planetary orbits, where r is the mean distance of the orbit from the sun, and T is the period of the orbit.

Kepler published the first two laws in 1609. He published all three laws in 1621 in his *Epitome of Copernican Astronomy*, which was Kepler's major publication. It was the first astronomical textbook based on the Copernican system, and was the primary source of information on the subject for 30 years.

Kepler died in poverty in 1630. Because of war, Kepler was unable to collect the arrears of his Imperial salary. Another personal blow to Kepler was that his mother was imprisoned 13 months for witchcraft, and died in 1622 soon after her release. [58]

Galileo's Measurements of Falling Bodies

Besides his pioneering investigation with the telescope, Galileo also performed important research describing the motions of falling bodies. The most well known was his demonstration that bodies of different weight fall at the same rate, provided that the force due to air resistance is negligible. However, this was only a small part of his research with falling bodies.

Galileo's measurements proved that a body falls with a constant acceleration, which we call the acceleration of gravity. The acceleration of gravity, which is denoted g, is equal to 9.8 meter/second per second, or 32.2 feet/sec per second. To simplify our calculations, this book approximates the acceleration of gravity g as 10 meter/sec per sec.

Figure 5-1 gives plots of the velocity and the distance dropped for a falling body, versus the time in seconds. The formulas for velocity and distance are as follows, where t is time in seconds:

Velocity: $V = gt = 10 t$ meters per second

Distance: $x = \frac{1}{2} g t^2 = 5 t^2$ meters

Acceleration is the rate of change of velocity. Diagram (a) shows that the rate of change of velocity (the slope of the velocity plot) is constant. This means that the acceleration is constant. Since the acceleration is 10 meter/sec per second, over every second the velocity increases by 10 meter/sec. The velocity is 10 meter/sec in one second, 20 meter/sec in 2 seconds, 30 meter/sec in 3 seconds, etc.

The distance dropped is proportional to the square of the time in seconds. In 3 seconds the body drops 45 meters (148 ft.). At that time the velocity is 30 meter/sec or 67 miles per hour.

It is often claimed that Galileo dropped objects from the leaning Tower of Pisa to prove that objects of different weight fall at the same rate, but this claim has been disputed. Insight into this issue can be gained by recognizing that this demonstration was only a small part of Galileo's experiments with falling objects. Galileo could not have implemented his much more complicated measurements of the acceleration of gravity from the Tower of Pisa. Consequently he may never have used the Tower of Pisa for any of his experiments with falling objects.

It makes a fanciful story to have Galileo dropping objects from the leaning Tower of Pisa, but it seems doubtful that this actually occurred.

[a]

[b]

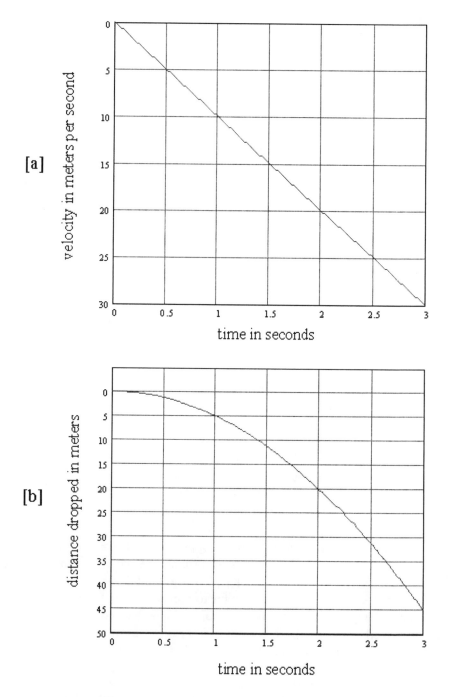

Figure 5-1: Velocity in meters per second and distance dropped in meters for a falling body, versus time in seconds

Calculation of the Acceleration of Gravity

Let us apply Newton's laws to calculate the acceleration of gravity, which is denoted g. In Table 5-1, item (a) gives the general formula for gravitational force between two bodies (1) and (2), which is specified by Newton's law (1), given earlier. Parameter M_1 is the mass of body (1), M_2 is the mass of body (2), and d_{12} is the distance between the centers of bodies (1) and (2). Parameter G is the constant of proportionality of the equation, which is called the *gravitational constant*.

Item (a) shows that the gravitational force is proportional to the product of the two masses (M_1 and M_2), and is inversely proportional to the square of the distance (d_{12}) between the two bodies.

In item (b), the general formula of item (a) is applied to obtain the weight (W) of an object. The weight is the gravitational force exerted by the earth on an object at the surface of the earth. Parameter M is the mass of the object, M_e is the mass of the earth, and r_e is the radius of the earth, which is the distance between the center of the earth and the center of the object.

Table 5-1: *Derivation of formula for acceleration of gravity*

Characteristic	Value
(a) General law of gravitational force	$GM_1M_2/(d_{12})^2$
(b) Weight (W) of object of mass M on earth	$GMM_e/(r_e)^2$
(c) Acceleration (g) of falling object (W/M)	$GM_e/(r_e)^2$

When this object is allowed to fall, the gravitational force W causes the object to accelerate downward with the acceleration of gravity (g). By Newton's law (2), the gravitational force W is equal to the mass M of the body multiplied by the acceleration of gravity (g). Hence, (g) is equal to the ratio (W/M), as shown in item (c). Dividing item (b) by M gives the formula for the acceleration of gravity shown in item (c).

The value for the acceleration of gravity (g) was known in Newton's time, and the radius of the earth (r_e) was known. Hence, Newton was able to calculate from the formula of item (c) the value of the product (GM_e). However, Newton had no direct means of finding the mass of the earth M_e, and so he could not determine the actual value for the gravitational constant G.

In 1798, Henry Cavendish measured the gravitational constant G

90   *The Scientific Story of Creation*

directly by sensing the gravitational attraction between two lead spheres. When G was known, the mass of the earth M_e could be calculated. Hence it was said that, "Cavendish weighed the earth in his experiment". The Cavendish experiment is described later in this chapter.

The Orbits of Planets around the Sun

Physical principles of a circular orbit. The gravitational pull of the sun on the earth holds the earth in its orbit as it revolves around the sun. In accordance with Newton's fourth law that for every action there is an equal and opposite reaction, the earth exerts the same force on the sun as the sun exerts on the earth. The gravitational pull of the earth on the sun moves the center of the sun slightly as the earth moves in its orbit around the sun.

You can illustrate the motion of the earth in its orbit by swinging a ball on the end of a string. You are the sun and the ball is the earth. The tension in the string represents the gravitational attraction between the sun and the earth. The force that the string applies to the ball is the gravitational force that holds the ball (earth) in a circular trajectory. If you let go of the string, the ball flies away in a straight line (except for the downward drop due to gravity). Thus the gravitational pull from the sun keeps the earth in its orbit, and prevents it from flying off into space.

The gravitational force that the sun applies to the earth causes the earth to accelerate. However this acceleration does not change the absolute value of the earth velocity (which we call the *speed*), it merely changes the direction of the earth velocity.

This concept is illustrated in Fig. 5-2. At time (1) the earth velocity is shown by the arrow V_1, which is called a "vector". The direction of the V_1 vector arrow shows the direction of the earth velocity at that instant, and the length of the vector arrow is proportional to the speed of the earth (30 km/sec).

At time (2) the earth velocity is shown by vector arrow V_2. The velocity vectors show the instantaneous values of velocity at points (1) and (2). The lengths of the two velocity vectors are the same, because the absolute value of the earth velocity (the *speed*) is constant. However, the velocity vector changes direction. Each velocity vector is tangent to the circular orbit at the instantaneous location of the earth.

In diagram (b), vectors V_1 and V_2 are moved without changing direction, so that they start at the same point. Vector ΔV is the difference between these two vectors, and is constructed by drawing a vector from the tip of vector V_1 to the tip of vector V_2. Vector ΔV is the amount by which the earth velocity changes between points (1) and (2).

The Greek letter Δ (delta) is commonly used to denote a difference. In this case ΔV is used to denote a difference in velocity V.

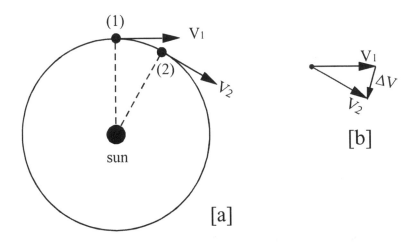

Figure 5-2: Change in the velocity of a planet as it rotates around the sun in a circular orbit

Acceleration is the change of velocity divided by the time interval over which the velocity change occurs. If we divide the length of the velocity difference vector ΔV by the time interval between points (1) and (2) we have the average acceleration between points (1) and (2). The earth accelerates between points (1) and (2) even though the speed of the earth stays constant.

This shows that velocity must be treated as a vector when acceleration effects are calculated. A vector has both direction and amplitude. If the acceleration produces a velocity change in the direction of the velocity vector, the speed changes. If the acceleration produces a velocity change perpendicular to the velocity vector (which is the case in Fig 5-2), the velocity direction changes but the speed does not.

Calculation of orbit parameters. Let us apply Newton's laws to calculate the path of a planet with a circular orbit. We will see that this agrees with Kepler's third law. In Table 5-2, item (a) shows the gravitational force between the sun and the earth, which is obtained from the general gravitational-force formula in item (a) of Table 5-1. Parameter M_e is the mass of the earth, M_s is the mass of the sun, and d_{se} is the distance between the centers of the sun and the earth.

92   The Scientific Story of Creation

Table 5-2: *Gravitational force on the earth that is required to hold the earth in its orbit around the sun*

Characteristic	Value
(a) Gravitational force between sun and earth	$GM_eM_s/(d_{se})^2$
(b) Acceleration of body in circular orbit of radius r	V^2/r
(c) Acceleration (A_e) of earth in its orbit	V^2/d_{se}
(d) Required acceleration force on earth (M_eA_e)	M_eV^2/d_{se}
(e) Required equality: ($V^2d_{se} = V^2r$)	GM_s

The *average acceleration* of the earth between points (1) and (2) is equal to the length of vector ΔV divided by the time interval between points (1) and (2). It can be shown with calculus that the *instantaneous acceleration* of a planet in a circular orbit is equal to (V^2/r), where V is the velocity of the planet in its orbit, and r is the radius of the orbit. This is shown in item (b) of Table 5-2. Since the earth orbit is essentially circular, the acceleration of the earth in its orbit is V^2/r, which is V^2/d_{se}, as shown in item (c). Parameter V is the velocity of the earth, and d_{se} is the radius (r) of the earth orbit, which is the distance between the sun and the earth.

To achieve the required acceleration A_e of the earth, the sun must exert a gravitational force on the earth equal to the earth's mass times its acceleration. This required gravitational force (M_eA_e) is shown in item (d). The acceleration force of item (d) is obtained by multiplying the earth acceleration of item (c) by the mass of the earth M_e.

Item (d) gives the acceleration force that must be exerted on the earth by the sun to hold the earth in its orbit and keep it from flying off into outer space. We set this acceleration force of item (d) equal to the gravitational force on the earth in item (a). This yields the equality of item (e), which shows that (V^2d_{se}) is equal to (GM_s). Since d_{se} is the radius r of the earth orbit, this requires that V^2r must be equal to (GM_s).

The expression (GM_s) is a constant for our solar system. Hence (V^2r) must be the same for all circular planetary orbits. Parameter r is the radius of the orbit and V is the velocity of the planet.

Kepler expressed his third law in terms of the period T of the orbit. Multiplying the velocity V of the planet by the period T gives the circumference of the orbit, which is 2πr, and so the velocity V is equal to 2πr/T. Therefore velocity V is proportional to r/T, and the expression (V^2r) is proportional to (r^3/T^2). Since (V^2r) is the same for all circular planetary orbits, (r^3/T^2) must also be the same. This result agrees with Kepler's third law, which was given earlier.

Thus we have seen that Kepler's third law is predicted by Newton's laws when we consider circular planetary orbits. If calculus is applied to Newton's laws, it can be shown that Newton's laws satisfy all three of Kepler's laws. This tells us that Newton's discovery of calculus was the key that allowed Newton to develop his laws of mechanics.

The accurate empirical laws developed by Kepler were essential elements in the development of Newton's theory. One of the greatest tragedies of scientific history was that Johannes Kepler died in poverty, despite the revolutionary character of his discoveries.

Orbit of the Moon around the Earth

To the orbit of the moon around the earth, Newton applied the same principles that he used to calculate the orbits of planets around the sun. From his analysis of the moon orbit, Newton proved that the following equality must hold:

$$V_m^2 d_{em} = g r_e^2$$

Parameter V_m is the velocity of the moon in its orbit, d_{em} is the distance from the earth to the moon, r_e is the radius of the earth, and g is the acceleration of gravity on earth. All of the quantities in this equation were known to Newton. Newton demonstrated that this equality was satisfied within measurement error and thereby provided strong evidence that his theory was valid. Newton withheld publication of his *Principia* until the latest measurements of distance to the moon and the diameter of the earth agreed with his theory.

Newton's laws have predicted with very high accuracy the motions of planets, moons, comets, and other bodies in our solar system. It was not until 1916 that Einstein proved with his General theory of Relativity that Newton's gravitational theory is not absolutely accurate. Einstein showed that relativistic effects produce a tiny error in the orbit of the planet Mercury that cannot be calculated from Newton's theory.

Engineering Use of Newton's Laws

By demonstrating that his laws accurately explain the motions of the planets and our moon, Newton established the validity of his laws. This proved that his laws also apply to the motions of objects here on earth, and so his laws became powerful engineering tools for practical applications. In the 1600's, advances in mechanical equipment had stimulated the search for basic scientific knowledge. Engineers needed to know how to build better mechanisms, and Newton's laws had immediate practical use in engineering applications.

The famous and brilliant German mathematician, philosopher, and statesman Gottfried Wilhelm Leibniz (1646-1716) invented calculus independently of Newton. He discovered calculus in 1675, nine years after Newton, but he published it in 1684 before Newton did. With the help of the Swiss mathematicians Jacques Bernoulli (1654-1705) and his brother Jean Bernoulli (1667-1748), the Leibniz calculus concepts were refined into a convenient scientific tool that became widely used.

It was the combination of this calculus tool developed by Leibniz and the Bernoulli brothers, along with Newton's laws, that gave the world a practical engineering approach to mechanics.

Another scientific achievement of Leibniz was his invention in 1672 of a calculating machine for multiplying, dividing and taking square roots.

Why Are Astronauts Weightless?

Why do astronauts in space experience weightlessness? Although this phenomenon is well known today, many people do not understand why it occurs.

Table 5-1 gives in item (c) the acceleration of gravity on the surface of the earth, which is $GM_e/(r_e)^2$. At the location of a space vehicle in orbit around the earth, the parameter r_e in this formula should be replaced by the radial distance r at that point from the center of the earth. Thus, the acceleration of gravity g is equal to GM_e/r^2, where r is equal to $(r_e + h)$. Parameter r_e is the radius of the earth (6378 km), and h is the altitude of the space vehicle.

The International Space Station is at an altitude of 418 km (260 miles). At this altitude the acceleration of gravity is 88 percent of the value at the surface of the earth.

Therefore, the gravitational force pulling the astronaut in the space vehicle toward the center of the earth is nearly the same as the weight that the astronaut experiences on the ground. Nevertheless, the astronaut is weightless in space. Why?

The reason for the weightlessness is that the space vehicle and the astronaut are in a free-fall condition. They are continually accelerating toward the center of the earth with an acceleration of gravity g corresponding to that altitude. However, the vehicle is traveling so fast that the altitude of the space vehicle stays constant. The acceleration merely changes the direction of the velocity vector. The acceleration of the satellite in a circular orbit is equal to V^2/r, where V is the satellite velocity. Setting V^2/r equal to the acceleration of gravity g, which is GM_e/r^2, shows that the International Space Station at its altitude of 418 km stays at constant altitude if its velocity V is 7.67 km/sec (4.77 mile/sec).

Thus the space vehicle and the astronaut are continually falling toward the center of the earth, even though their altitude does not change. In this free-fall condition, the astronaut is weightless.

A sky-diving parachutist experiences weightlessness for a few seconds after jumping out of an airplane, until the velocity is sufficient for air resistance to offset the force of gravity. The wind force from the air is proportional to the square of velocity. It matches the sky diver weight at a speed of about 120 mph when the sky diver's body is oriented horizontally, or 180 mph when the body is oriented vertically. During the first few seconds, the wind force is very small, and so the falling sky diver is essentially weightless.

A simple way to experience temporary weightlessness is to jump from a high platform into water.

How Cavendish Weighed the Earth

In 1798 Henry Cavendish (1731-1810) measured the gravitational constant G by sensing the gravitational attraction between lead spheres. Henry Cavendish was a brilliant and wealthy scientist, but was reclusive and eccentric. He is best known for his chemistry research, which included the discovery that water is a compound of hydrogen and oxygen.

The experiment by Henry Cavendish is shown in Fig. 5-3. A horizontal dumbbell, with lead balls at each end, was suspended from a thin vertical wire. A pair of much larger lead spheres was mounted on a rotatable structure, which moved the spheres just outside the lead balls. As the large spheres were moved past the balls of the dumbbell, the gravitational attraction between the dumbbell balls and the large spheres made the dumbbell rotate, and caused the wire to twist slightly. The amount of twist was sensed optically to give a measure of the gravitational attraction.

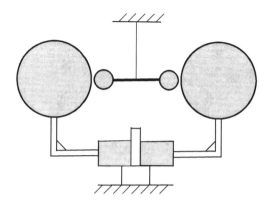

Figure 5-3: The Cavendish experiment to measure Newton's gravitational constant G

The weight of a dumbbell ball was about 10 million times greater than the tiny gravitational force between the ball and the large sphere. Nevertheless the experiment was sufficiently sensitive to detect this gravitational force. From this experiment, Cavendish measured the gravitational constant G within an error of 1.4 percent.

Table 5-1 showed that the acceleration of gravity g is equal to GM_e/r_e^2, where M_e and r_e are the mass and radius of the earth. Since the values of g and r_e were known, this measurement of the gravitational constant G allowed Cavendish to calculate the mass of the earth M_e. Consequently it was said that, *"Henry Cavendish weighed the earth in his experiment"*. He found that the average density of the earth is 5.5 times that of water.

Coordinates to Specify a Vector

In order to apply Newton's laws in a general manner, one needs a coordinate system to specify a vector. A point in three-dimensional space is usually described either in terms of *rectangular coordinates* or *spherical coordinates*. Let us examine these two coordinate systems.

Rectangular coordinates. A map allows one to specify a location in two dimensions. One can describe a location on a map by giving the easterly and northerly distances measured from a set of north-south and east-west reference lines, which are called axes.

One can specify the location of an airplane relative to a map by giving the east-west and north-south coordinates of the point directly under the aircraft, plus the vertical position of the aircraft, which is usually given as the altitude above mean sea level. This gives a three-dimensional specification of the aircraft location in *rectangular coordinates*.

Figure 5-4 shows a general method for describing a location in three-dimensional *"rectangular coordinates"*. These are also called *"Cartesian Coordinates"*, because they were first used by the French mathematician, Rene Descartes (1596-1650).

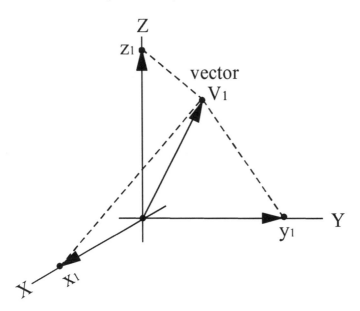

Figure 5-4: Rectangular coordinates of a vector

In Fig. 5-4, a location is specified in terms of three reference axes, labeled x, y, and z. It is often convenient to regard the z-axis as being vertical, where the x, y axes lie in the horizontal plane, with the x-axis pointing east and the y-axis pointing north. Although the x, y, z axes often represent *East* (x), *North* (y), and *Vertical* (z), any orientation of the three axes can be used provided that their relative orientation is preserved.

The arrow V_1 in Fig 5-4 is called a vector. A vector can represent any quantity that has direction as well as amplitude. For example, the vector might represent the position of an aircraft or its velocity. The vector is specified by giving the coordinates of the tip of the vector relative to the x, y, z axes. If the vector represents aircraft velocity, the coordinates of the vector are the components of aircraft velocity in the easterly, northerly, and vertical directions.

These coordinates are obtained by drawing the three dashed lines from the tip of the vector, perpendicularly to the x, y, z axes. The point at which a dashed line hits the x-axis is the x-component of the vector. The y-component and z-component are defined in a similar fashion. In this manner, one can uniquely specify any 3-dimensional vector in rectangular (or *Cartesian*) coordinates.

Spherical coordinates are another way to specify a vector. Spherical coordinates are related to the latitude-longitude coordinates for locating a point on the curved surface of the earth. Over a small region, the surface of the earth is approximately flat, and so a point can be located on a flat map. However, to specify location in an absolute manner relative to the spherical earth, one must consider the curvature of the earth and apply the angular measurements of latitude and longitude.

Latitude and longitude are defined in Fig. 5-5. For a point in the northern hemisphere, *north latitude* is measured in degrees from the equator in the northerly direction to the point of interest. For a point in the southern hemisphere, *south latitude* is measured in degrees from the equator in the southerly direction.

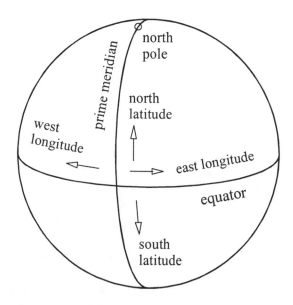

Figure 5-5: Latitude-longitude coordinates for specifying a location on the earth

A *meridian* is a north-south line of constant longitude, which is a semicircle drawn from pole to pole. The *prime meridian* (or *Greenwich meridian*) is the line of constant longitude that passes through the old Royal Observatory in Greenwich, England. The prime meridian is defined as the line of zero longitude.

A point that is east of the prime meridian is specified by its *east-longitude* coordinate. East longitude is the angle measured in degrees along the equator in the easterly direction from the prime meridian to the meridian that passes through the point. Similarly, a point that is west of the prime meridian is specified by its *west-longitude* coordinate. With latitude and longitude angles one can locate a point on the curved surface of the earth in two dimensions.

This latitude-longitude technique can be generalized to obtain a three-dimensional specification by including the radial distance measured from the center of the earth. For example, the location of a satellite circling the earth can be specified in the following manner. Consider a line drawn from the center of the earth to the satellite. The length of this line is the radial coordinate r of the satellite. The latitude and longitude coordinates are measured at the point where the line passes through the surface of the earth. The distance coordinate r and the latitude-longitude angular coordinates give a three-dimensional specification of the satellite location.

Relativity theory commonly uses both rectangular and spherical coordinates to specify spatial information. To locate a point in three dimensions with rectangular coordinates, three distances are specified, which are measured along the x, y, and z axes. With spherical coordinates, the radial distance from the origin to the point of interest is specified, along with two angles, which are similar to the latitude and longitude coordinates for locating a point on the curved surface of the earth.

Chapter 6

The Nature of Light

The theory of Relativity evolved from a study of the speed of light. To introduce that theory, this chapter explains the physical principles of light propagation and reviews the history of our knowledge of light.

What is a Light Wave?

We know that light and sound are waves. We can help to understand these waves by examining the propagation of a wave on water.

Mechanical Waves

We can envision a wave by dropping a stone into a smooth pond. The water waves propagate outward from the point where the stone hits the water. An individual water particle oscillates back-and-forth and up-and-down, following an elliptical path. It is the energy of the wave that propagates across the pond, not the water itself. This effect can be observed by noticing that an object floating on the surface moves very little as the wave passes it.

If we drop two stones into the pond at once, two sets of waves are produced, and these two sets of waves interfere with one another. Similar wave interference occurs with sound and light waves. Figure 6-3, on page 110, will explain a famous light experiment involving wave interference, which was used to prove that light propagates as a wave.

Like a wave on water, sound is a mechanical vibration. One can feel the vibration in a musical instrument when a tone is produced. The instrument vibrates the air, producing sound that travels through the air to the ear that hears it. Sound is a compression wave, in which the air particles vibrate back and forth in the direction of propagation of the sound.

Electromagnetic Waves

A wave on water and a sound wave are mechanical waves, which propagate by vibrating a medium. Although a light wave is similar in certain respects to these mechanical waves, it is fundamentally very different. A light wave is a packet of oscillating electric and magnetic fields, and so is called an electromagnetic wave.

Types of electromagnetic waves. A light wave is the same as a radio wave except that it oscillates at a much higher frequency. Table 6-1 lists the frequencies and wavelengths of typical classes of electromagnetic waves over the electromagnetic spectrum. All of these waves are physically the same. They differ only in frequency.

Table 6-1: *Approximate frequencies and wavelengths of typical classes of electromagnetic waves*

signal type	frequency	wavelength	frequency ratio
Navigation	10 kHz	30 kilometer	0.01
AM Radio	1 MHz	300 meter	1.0
Television	500 MHz	60 centimeter	500
Precision Radar	10,000 MHz	3 centimeter	10 thousand
Visible Light		0.6 micrometer	500 million
X-Rays		0.1 nanometer	3,000 billion

Standard AM (amplitude modulation) radio operates at a frequency of about 1 megahertz (MHz). One hertz (Hz) means one cycle per second, and so 1 MHz means 1 million cycles per second. Television operates at a frequency of about 500 megahertz. Precision radar operates at a frequency of about 10,000 megahertz.

At 10,000 MHz, the wavelength of the electromagnetic wave is 3 centimeters. Wavelength is equal to the ratio c/f, where f is the frequency and c is the speed of light (300 million meter/sec). For frequencies higher than that of radar, it is usually convenient to deal with wavelength rather than frequency.

Precision radar has a wavelength of about 3 centimeters, and visible light has a wavelength of about 0.6 micrometer, where one micrometer is one millionth of a meter, or one-thousandth of a millimeter. An X-ray wave has a wavelength of about 0.1 nanometer. One nanometer is one-billionth of a meter, or one thousandth of a micrometer.

The last column of Table 6-1 shows the ratios of the frequencies of the various waves relative to that of an AM radio wave. The frequency of television is 500 times greater than AM radio; the frequency of precision radar is 10 thousand times greater; the frequency of light is 500 million times greater; and the frequency of x-rays is 3 trillion times greater. (One trillion is 1000 billion.)

Radio frequencies that are much lower that that of AM radio (1 MHz) are also employed. A frequency of 10 kilohertz (10,000 cycles per second) is used for navigation. This has a wavelength of 30 kilometers (km). The United States military uses a frequency of about 100 Hz for communication with its submarines. This has a wavelength of 3,000 km. This "extra-low-frequency" wave has the advantage that it can penetrate water to a great depth.

Meaning of Electric and Magnetic Fields

An electromagnetic wave consists of oscillating electric and magnetic fields. To help understand the electromagnetic wave, let us consider some examples of electric and magnetic fields.

We observe the effect of an electric field in a thunderstorm. The severe winds of the storm remove electrons from the clouds and deposit them on the ground. The clouds become charged positively with respect to the ground, and a large electric field builds up between the clouds and the ground. Eventually this electric field gets so strong that an electric spark (called lightning) jumps between a cloud and the ground. Electrons flow from ground to the cloud, and temporarily eliminate the electric field.

Our earth has a magnetic field, which we can sense with a magnetic compass. A magnetic compass is a magnet that is free to turn and points in the direction of the magnetic field. Our earth acts like a huge magnet, which has north and south poles that are close to the poles about which the earth rotates, and so a magnetic compass points approximately in the direction of true north.

Although electric and magnetic fields are quite different, they are closely related, as shown by the following:

(1) A changing magnetic field produces an electric field
(2) A changing electric field produces a magnetic field

Property (1) can be observed in an automobile generator, which delivers electric current to keep the battery charged. The rotating part of the generator, called the rotor, acts as a magnet. As the rotor revolves, its magnetic field moves through a coil of wire wound on the fixed part of the generator (the stator). This action produces an electric field, which generates an electric current to charge the battery. *This shows that a changing magnetic field produces an electric field.*

Property (2) can be observed in an electromagnet. When electrons flow they produce an electric current, which represents a moving electric field. If the current is fed through a coil of wire, an electromagnet is formed which generates a magnetic field. The current in the coil of wire represents a moving (or changing) electric field, and this changing electric field produces a magnetic field. *Hence, in accordance with property (2), a changing electric field produces a magnetic field.*

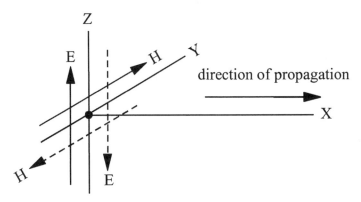

Figure 6-1: Oscillating electric and magnetic fields that form an electromagnetic wave.

Principle of an Electromagnetic Wave

An electromagnetic wave consists of oscillating (or changing) electric and magnetic fields. According to property (1), the oscillating (or changing) magnetic field produces an oscillating electric field, and, according to property (2), the oscillating (or changing) electric field produces an oscillating magnetic field.

An electromagnetic wave is illustrated in Fig. 6-1. Electric and magnetic fields oscillate at right angles to one another. The oscillating electric field produces the oscillating magnetic field, and the oscillating magnetic field produces the oscillating electric field. These two oscillating fields support one another, and thereby form a wave that propagates at the speed of light.

The wave propagates in a direction that is perpendicular to the plane within which the electric and magnetic fields oscillate. The electric field E vibrates up and down in the direction of the z-axis. The magnetic field H vibrates back and forth in the direction of the y-axis. The solid E arrow shows the direction of the electric field during one half cycle, and the dashed E arrow shows the direction in the next half cycle. Similarly, the solid and dashed H arrows show the directions of the magnetic field in alternate half cycles.

The oscillating electric and magnetic fields are out of phase with one another, so that when one field is a maximum the other is zero. The oscillating electric and magnetic fields support one another, to form an electromagnetic wave that propagates in the x-direction. The y-z plane (within which the electric and magnetic fields oscillate) is perpendicular to the x-direction in which the wave propagates.

Light is one type of electromagnetic wave. Light is a packet of electric and magnetic fields that travel at the speed of light. Light acts like a wave, because the electric and magnetic fields oscillate as they propagate. However, light is not like a sound wave or a wave on water, which propagate by vibrating a medium. There is no medium involved in the propagation of light. Light can travel through a vacuum, but sound cannot because there is no medium in the vacuum to vibrate.

Early Concepts of Light

Since the days of the Classical Greeks we have known that sound is a vibration, or oscillation. One can feel the vibration in a musical instrument. The sound vibration travels as a wave through the device that generates the sound, and then through the air to the ear that receives it. Sound moves through the air as a compression wave. The air particles vibrate back and forth in the direction of propagation of the sound wave.

But what is light? Is it a wave like sound? Or is it a stream of particles shot from the light source like stones from a slingshot?

Galileo and Kepler

The science of optics started with the making of spectacles to correct visual defects. The first documented use of eyeglasses was a description by Roger Bacon in 1268 of a hand-held magnifying glass, but the Chinese may have used eyeglasses much earlier. The first eyeglasses were made of quartz. By 1629 eyeglass making was so advanced that a guild of spectacle makers was formed in England.

About 1608 a Dutch spectacle maker, Hans Lippershay or Jacob Metius, built the first telescope, which had a magnification of 3 to 4. It consisted of a convex objective lens and a concave eyepiece. (A convex lens is curved outward, whereas a concave lens is curved inward, like a *cave*.) Galileo heard of this and began making better telescopes. In 1609 he supplied the governor *(doge)* of Venice with an 8-power telescope. This was so valuable for naval use that Galileo's salary was doubled, and he received lifelong tenure as a professor. Later in that year, Galileo built a 20-power telescope, which he directed at the heavens.

In 1610 Galileo revolutionized astronomy with his telescope. He studied our moon and found that it has mountains and valleys like the earth. He looked at Jupiter and made the fantastic discovery that Jupiter has moons of its own. He saw that Venus has phases like our moon, which showed that Venus must revolve around the sun. A problem with Galileo's telescope was that the field of view was very narrow.

Kepler invented the astronomical telescope, which uses a convex rather than a concave lens for the eyepiece. This telescope has a wider field of view, but has an inverted image and so is not good for terrestrial use. The inverted image is not a problem in astronomy.

Modern high-quality terrestrial telescopes and binoculars use the Kepler telescope concept, along with an additional lens or prism to compensate for the inverted image. Inexpensive *field-glass* binoculars use the simple Galileo telescope concept with a concave eyepiece.

The Optical Discoveries of Isaac Newton

Many people followed Galileo's lead and built telescopes to study the heavens. A telescope of that time had poor resolution because of chromatic aberration. The light from its simple lens does not focus at a single point, because the wavelengths of light are refracted at different angles. This produces colored fringes around the image.

Modern refracting optical instruments, such as telescopes and cameras, use achromatic lenses, which have two kinds of glass with compensating refraction characteristics. Refraction differences over the visible wavelength band are cancelled, and so chromatic aberration, with its color fringes, is eliminated. The achromatic lens was invented in 1757 by the British optician John Dollond.

Achromatic lenses were not available in the 1600's, and so refracting telescopes at that time were poor. In 1668 Isaac Newton made a tremendous advance in telescope design by fabricating a spherical reflecting mirror from metal, and he built from this the first reflecting telescope. All wavelengths are reflected from a mirror at the same angle, and so the primary (or "objective") optical element of his reflecting telescope had no chromatic aberration. The eyepiece used a refracting lens, but chromatic aberration from the eyepiece has much less effect than that from the objective lens. This reflecting telescope was a tremendous improvement, and it made Newton famous.

The spherical mirror in Newton's reflecting telescope was made from speculum metal, which is an alloy of tin and copper. Speculum metal was used to make reflecting telescopes until the late 1800's, when it was replaced by glass with a silver-coated surface.

Newton still wanted to build a refracting lens without chromatic aberration. He performed optical experiments with prisms in order to understand the nature of chromatic effects, so that he could compensate for chromatic aberration. He did not have glass with the proper characteristics to achieve this, and concluded that an achromatic refracting lens could not be built. However in his experiments he made a fundamental discovery that revolutionized the knowledge of light.

A normal light source, such as a candle or the sun, has a broad spectrum covering the visible band. We call this a *white* light, because it looks white to the eye. When a beam of white light is passed through a prism, the wavelengths contained in the white light are refracted at different angles, and thereby form a spectral pattern displaying the colors of a rainbow. It is obvious to us today that a prism separates white light into its component wavelengths, but this concept was not understood at all in Newton's day.

The problem was that the only instrument for detecting optical wavelength effects in Newton's time was the color vision of the eye. Color vision is a very complicated phenomenon, which is not well understood even today. We perceive sound as a succession of tones of different frequencies, but our perception of color is entirely different.

An important aspect of color perception is that vision strongly modifies the optical information that it receives. By means of adaptation processes, the eye compensates to a large extent for the illumination, and so the color of an object appears to be almost independent of the spectrum and intensity of the light that illuminates the object.

Our perception of color has four basic chromatic sensations: red, green, yellow, and blue. It also has the achromatic color sensations, white, gray and black. The four chromatic sensations are arranged in pairs: red-green and yellow-blue. We perceive either red or green, but not a combination of red and green; and we perceive either yellow or blue, but not a combination of yellow and blue.

Therefore we can represent our chromatic sensations in terms of two perpendicular axes, as shown in diagram [a] of Fig 6-2, where green is the negative of red, and blue is the negative of yellow. The chromatic sensation is zero at the origin, and so at that point we perceive the achromatic sensation, which may be white, gray, or black.

The figure also shows the intermediate chromatic sensations, which are combinations of red, yellow, green, and blue. A combination of red and yellow sensations produces orange; a combination of yellow and green produces yellow-green (olive); a combination of blue and green produces blue-green (aqua); and a combination of blue and red produces violet and purple. This shows that the chromatic color sensations form a circle.

When artists arrange colors into an orderly array, they generally space the colors around the color circle in a different manner. A common form of the artist's color circle is shown in diagram [b]. The six colors (red, orange, yellow, green, blue, and violet) are equally spaced around the color circle. This spacing has the advantage that colors opposite one another in the color circle are "contrasting" or "complementary" colors.

When light is projected through a prism, it forms a linear pattern with the colors of a rainbow, as shown in diagram [c]. The chromatic colors are now arranged in the linear sequence: red, orange, yellow, green, blue, and violet. How do we relate this linear pattern of colors projected from a prism to the circular pattern of colors in the artist's color circle?

A reader who has never studied the issue of color perception may feel confused. The sense of color that he has taken for granted suddenly has become an enigma. Yet his confusion is mild in comparison to what was experienced by the associates of Newton in the Royal Society.

Newton had discovered that white light is a combination of different kinds of light of different "refrangibility", which are distinct from one another. However, he did not know that they were of different wavelengths. When he presented his results to the Royal Society in 1672, they were flatly rejected.

The president of the Royal Society, Robert Hooke, had his own theory of color. Hooke believed that light is like a pulse that changes in shape when reflected from a colored object. Hooke was sure that his theory gave a much better answer to why "yellow mixed with blue produces green" than did Newton's explanation.

As reported by Cohen [61], Newton published several detailed responses to arguments against his optics research in the *Transactions of the Royal Society*, but was unable to get anyone to even try to repeat his experiments. Finally Newton gave up in disgust, and concluded that: *"A man must either resolve to put out nothing new, or to become a slave to defend it"*.

Newton put his optics research aside to study planetary motions. His theory of gravity was published in 1687 as *Philosophae Naturalis Principia Mathematica*. Newton's *Principia* gave him great fame, and in 1703 he was elected president of the Royal Society. In 1704 Newton published his book *Opticks* [63], which was written in English, and described his research with prisms. Nearly all of *Opticks* had been written 30 years earlier.

Newton's *Opticks* revolutionized the knowledge of light. Nevertheless, readers of this book in later years have generally misunderstood the meaning it had when it was written. *The basic message of Newton's Opticks was that white light is a combination of different rays having unique refraction characteristics. This finding is so obvious to us today; it is hard for us to realize that it was a revolutionary concept in Newton's time.*

The Wave Property of Light

Newton was not particularly concerned about the question of whether a light beam was a wave or a stream of particles (or "corpuscles"). However the wave concept seemed dubious. He knew that sound travels through the air as a compression wave by vibrating the air, but sound cannot be transmitted where there is no air.

What is this mysterious medium, called the *luminiferous aether*, that is vibrated by light and allows a light wave to be transmitted over infinite distances in empty space? How can the aether be so thin that it offers no resistance to planets as they travel in their orbits, yet be so stiff that it transmits light vibrations at a fantastically high speed? (The approximate speed of light was known in Newton's time).

Consequently Newton assumed that a light beam was a stream of particles that he called "corpuscles". He postulated that the light rays of different "refrangibility" that make up a beam of white light are corpuscles of different "size". When a beam of white light is passed through a prism, the light corpuscles of different size follow separate paths, and thereby are spread out to form a colored spectrum.

Christiaan Huygens (1629-1695) was a famous Dutch mathematician who lived at the time of Newton. He believed in the wave concept of light, and made extensive mathematical analyses of light-wave propagation.

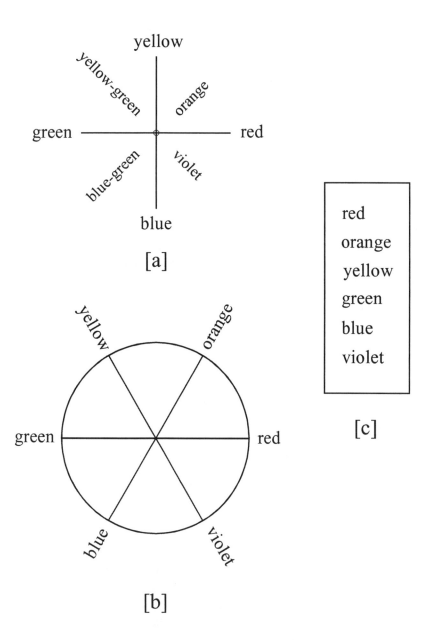

Figure 6-2: Complexity of color perception, which confused Newton's optical research; [a] coordinates of chromatic color perception; [b] the artist color circle; [c] linear rainbow array of colors projected by a prism

Because Newton was so famous, his postulate that light is a stream of corpuscles was generally assumed to be correct during the 1700's. However in 1802 Thomas Young (1773-1829) challenged this concept in a Royal Society paper. Young used the theoretical work of Huygens to assist him in his research.

Thomas Young was a physician, but his primary contributions were in other areas. He did studies on the elasticity of materials, and his name is well known today by the term *Young's modulus of elasticity*. He worked on the translation of Egyptian Hieroglyphics from the Rosetta Stone, which carries an inscription written in Hieroglyphics, Greek, and Demotic, a simplified script that is related to Hieroglyphics. Deciphering the Rosetta Stone was the key that allowed archaeologists to read the mysterious writings found in Egyptian tombs. Thomas Young achieved the initial translations, and discovered that at least some Hieroglyphics are phonetic symbols. Jean Champollion built on Young's work, and is usually given the credit for deciphering the Rosetta Stone.

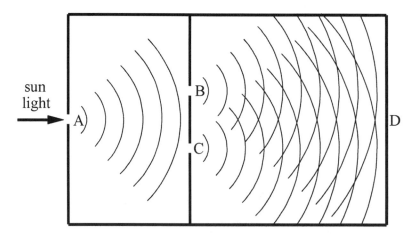

Figure 6-3: Light interference experiment by Thomas Young

However the primary achievement of Thomas Young was to demonstrate the wave nature of light with the experiment shown in Fig. 6-3. Young passed sunlight through a pinhole (A). The light was refracted by (A), and diverged so that it reached pinholes (B) and (C). The light was refracted again at pinholes (B) and (C). The diverging light beams from (B) and (C) interfered with one another, to produce an interference pattern that was projected onto surface (D). (This experiment is often performed with slits, but Young used pinholes.)

This optical interference pattern proved that light is a wave. With this experiment Young was able to measure the wavelength of light.

From 1802 to 1804 Young published his light-interference experiments in the *Transactions of the Royal Society*. The response was a merciless attack against him, which included two anonymous discussions of Young's work in the *Edinburgh Review*, probably written by Lord Brougham. As shown in the following [64], Young was ridiculed for being so arrogant as to question the great Sir Isaac Newton:

"We wish to raise one feeble voice against innovations that can have no other effect than to check the progress of science, and renew all those wild phantoms of the imagination which Bacon and Newton put in flight from her temple. The paper contains nothing that deserves the name of either experiment or discovery."

A few years later a young French physicist, Augustin Fresnel (1788-1827), who was unaware of Young's work, began to perform similar light interference experiments. His friend, Dominique Arago, brought him in contact with Thomas Young. In 1819 Fresnel won the Scientific Competition Award by the French *Academie des Sciences* for his research on light-wave interference. This involved a precise experiment combined with sophisticated mathematical analysis. This award placed the wave theory of light on a solid footing, but opposition to the theory continued for years.

Augustin Fresnel died in 1827 at age 39, and the world lost a great scientist. He had been sickly for years with tuberculosis.

Newton has often been criticized for insisting on the corpuscular theory of light, but this criticism is unjustified. When Newton described his optical experiments, he did not particularly care whether light was a wave of a stream of particles. His "rays of different refrangibility" could just as well be *waves of different wavelength or particles of different size*. He needed a theoretical model to explain his experiments, and he picked the particle ("corpuscle") concept because it seemed more believable.

Because of the stubbornness of Robert Hooke, president of the Royal Society when Newton did his optical research, and the sheep-like mentality of others in the Society, Newton's optical findings were rejected. Nearly all of Newton's *Opticks* was written at that time, but it was not published until 30 years later. After publication, many people immediately repeated and extended Newton's experiments, and the science of optics advanced rapidly.

Newton's revolutionary idea was that white light is a combination of unique rays having different refraction characteristics. After a few years this idea became absurdly obvious, and so readers of Newton's *Opticks* gradually missed its real point. Instead many came to assume incorrectly that this document was written as a treatise expounding the corpuscular theory of light.

Because of Newton's enormous prestige, his casual corpuscular postulate became enshrined as absolute truth. Thomas Young faced strong opposition to his demonstration of the wave nature of light because it conflicted with the preconceived beliefs of many of his contemporaries. Young's research did not conflict in a substantive way with the statements made by Newton in his book *Opticks*.

Thus Thomas Young and Augustin Fresnel proved that light is a wave. But how does it propagate? The mysterious aether medium that transmits light waves was as enigmatic as ever.

The Electromagnetic Wave Concept

During the 1800's great advances were made in our knowledge of electricity and magnetism, from research performed by Coulomb, Volta, Oersted, Gauss, Weber, Ampere, Faraday, and others. These electrical and magnetic findings were tied together in 1873 by James Clerk Maxwell (1831-1879), and presented as a set of *electro-magnetic* equations. These equations deal with electrical and magnetic "fields", and so are called *electromagnetic field equations*.

Maxwell predicted from his equations that an electromagnetic "radio" wave can be generated, consisting of oscillating electric and magnetic fields. The predicted speed of this wave (based on measured electrical and magnetic parameters) was the same as the measured speed of light. Therefore Maxwell concluded that light must be an electro-magnetic wave of very high frequency.

In 1888 Heinrich Hertz (1857-1894) generated the electromagnetic "radio" wave that Maxwell had predicted. However it was not until 1895 that Guglielmo Marconi (1874-1937) achieved practical communication with radio over a few kilometers, using a radio wave of much lower frequency than Hertz. (*Guglielmo* is pronounced *"Gulliamo"* and is the Italian version of *William*.)

Marconi performed his first radio experiments in 1890 when he was only 16. He established telegraphic radio communication links across the English Channel in 1899, from England to St. John's, Newfoundland in

1901, and from England to Cape Cod, Massachusetts in 1903. His antenna on Cape Cod was supported on four enormous towers, 200 feet tall. The antenna consisted of wires that formed a conical mesh that rose from the building below to the top of the towers.

Maxwell's electromagnetic field equations are used extensively today in complicated analyses. They are the foundation that allows engineers to design many electrical devices, such as the antennas for transmitting and receiving radio, television, and radar signals. Marconi must have studied Maxwell's equations intensely to develop his very sophisticated and large antenna designs, which were the first antennas for transmitting and receiving radio signals. A difficult problem with radio communication in those days was that the device for detecting the radio signal had very poor sensitivity.

A light wave propagates by means of oscillating electric and magnetic fields, not by vibrating a mysterious substance called the *luminiferous aether*. Maxwell's electromagnetic field equations eliminated the need for an aether in light propagation, thereby removing the stumbling block that had caused Newton to reject the concept that light is a wave. Nevertheless Maxwell continued to assume in his writings that the aether still existed.

Search for the Velocity of the Luminiferous Aether

Why did Maxwell maintain the aether concept, even though the aether made no sense physically, and was not needed in the propagation of an electromagnetic wave? The following gives the probable explanation.

A sound wave travels with respect to the air, which vibrates, and so the velocity of the sound wave is measured relative to the air. Remember that sound cannot propagate where there is no air.

What happens with a light wave? Suppose that the light emitter and light receiver are moving relative to one another at a velocity that is an appreciable fraction of the speed of light. To what reference is the speed of light measured? Is the light speed measured relative to the emitter; or is it measured relative to the receiver; or (like sound) is it measured relative to the aether medium through which the light is propagating?

Figure 6-4 helps to clarify this issue. Diagram (a) illustrates the propagation of a sound wave. The sound wave propagates relative to the air at a velocity V_{sound} of 340 meter/sec. The relative velocity of the sound with respect to the emitter (or receiver) depends on the velocity of the emitter (or receiver) relative to the air.

For example, if the emitter moves relative to the air at velocity V_{em} in the direction of the sound wave, the speed of the sound relative to the emitter would be ($V_{sound} - V_{em}$). Similarly, if the receiver moves relative to the air at the velocity V_{rec} opposite to the direction of the sound wave, the speed of the sound relative to the receiver would be ($V_{sound} + V_{rec}$).

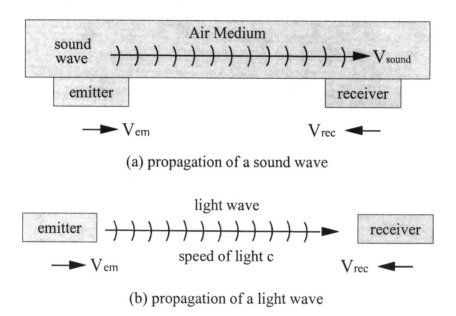

Figure 6-4: Propagation of light wave compared with propagation of sound wave

Now let us consider a light wave. Since light does not propagate relative a medium, it is not clear what we mean by the "speed of light". Let us assume that light moves at a fixed velocity c relative to the emitter. If the emitter is moving in the direction of the light wave with velocity V_{em}, as shown in diagram (b), the absolute velocity of the light would be ($c + V_{em}$). If the receiver is moving in opposition to the light wave with a velocity V_{rec}, the relative velocity between the receiver and the light should be equal to ($c + V_{em} + V_{rec}$).

This tells us that at the receiver the light should have a relative velocity equal to the nominal speed of light c plus the relative velocity between the receiver and the emitter.

But this cannot be true. If it were, the speed of light would be radically different for light received from different stars. We can tell from the Doppler wavelength shift that some stars are moving at very large velocities relative to earth. Nevertheless the speed of light does not vary measurably from star to star.

Such reasoning led most scientists to believe that light must move at a constant velocity relative to the aether, just as sound moves at a constant velocity relative to the air. Although sound propagates by vibrating the air, light does not propagate by vibrating the aether. ***Nevertheless, it was still believed that the aether provides an absolute reference medium, relative to which the light wave moves.***

The Michelson-Morley Experiment

This reasoning is apparently what led Maxwell to include the aether concept in his writings. This hypothesis is supported by the fact that Maxwell's influence was a major factor leading Michelson and Morley to perform the famous Michelson-Morley experiment for measuring the velocity of the aether.

The earth revolves around the sun at a velocity of 30 km/sec, which is 0.01 percent of the speed of light (300,000 km/sec). During the year the earth moves at ±0.01 percent of the speed of light in different directions, and so should move at ±0.01 percent of the speed of light relative to the aether.

Michelson and Morley built a very accurate instrument, using wave interference effects, that could measure differences in the speed of light in perpendicular directions with accuracy much better than 0.01 percent of the speed of light. The instrument was slowly rotated so that a given path was sometimes in the direction of the earth's motion, and sometimes perpendicular to that motion.

The instrument was applied many times during the year, but there was no detectable variation of the speed of light, regardless of the direction of the earth's motion. The results of the Michelson-Morley experiment were completely negative.

Many hypotheses were proposed to explain the negative results of the Michelson-Morley experiment. It was suggested that in the vicinity of the earth the aether moves with the earth, just like the earth's atmosphere. However, if this were true, the aether in various parts of our solar system would have different velocities, and so the speed of light would be different. If this occurred, the variation of the speed of light would be obvious.

The Contraction Hypothesis

Finally, Professors George FitzGerald and Hendrik Lorentz (1853-1928) independently proposed the hypothesis that the length of an object contracts when the object moves relative to the aether. A dimension in the direction of the relative velocity would contract by the following factor K:

$$K = \sqrt{[1 - (V/c)^2]}$$

where V is the velocity of the object relative to the aether, and c is the speed of light. As the Michelson-Morley instrument rotates, the two arms would change in length according to this formula, and this postulate can explain the negative results of the experiment. [8]

This contraction effect can be represented in a geometric manner as shown in Fig. 6-5. The hypotenuse AB of the right triangle has a length of unity. The length of the side BC is equal to the velocity ratio V/c. The third side AB is equal to the contraction factor K.

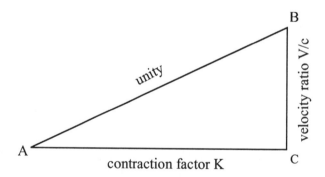

Figure 6-5: Right triangle showing in a geometric manner the relation between the V/c velocity ratio and the Lorentz contraction factor K

The Lorentz Transformation

Other experiments were performed to measure motion relative to the aether, which also had negative results. Lorentz found that he could explain all of these experiments by generalizing his contraction hypothesis. He developed a set of equations showing that time as well as distance measurements may be altered when a body moves relative to the aether. He proved that Maxwell's electromagnetic field equations would be invariant if modified by his equations. These Lorentz equations were published in 1904, and later became known as the *Lorentz transformation equations.* [8]

Since Maxwell's equations are invariant when modified by the Lorentz transformation equations, Lorentz concluded that it would probably be impossible to measure the velocity of a body relative to the aether if these transformation equations are valid.

But if this is true, how can the aether have any meaning? If it is impossible to detect the presence of this mysterious medium, why should we believe that the aether actually exists? Into this confusion stepped the brilliant young physicist, Albert Einstein. In 1905 Einstein [9] published his paper on Relativity the year after the Lorentz paper.

The Einstein Relativity Principle

Albert Einstein (1879-1955) approached the enigma of the speed of light by applying fundamental reasoning. He concluded that there is no such thing as *absolute velocity*; there is only *relative velocity*. **We can specify the relative velocity between two bodies, but not the absolute velocity of either one.**

Based on this principle, Einstein concluded that the aether cannot exist. Since there is no such thing as absolute velocity, there cannot be an aether medium that establishes an absolute reference frame for specifying velocity.

How do we specify the speed of light? Let us refer back to diagram (b) of Fig. 6-4. The light travels at the speed of light from the emitter to the receiver. Since the speed of light can only be specified in a relative sense, we can consider (1) that the speed of light is the relative velocity between the emitter and the light, and (2) that the speed of light is the relative velocity between the receiver and the light.

Since there is no such thing as absolute velocity, the velocity of an observer located at the emitter is no more (nor no less) preferred than the velocity of an observer located at the receiver. Consequently, both observers must measure the same relative velocity for the speed of light. An observer located at the emitter must measure exactly the same speed of light as an observer located at the receiver.

Thus, Einstein established the following principle: **Two observers moving at different velocities must measure exactly the same value for the speed of light, regardless of the relative velocity between them.**

Einstein then asked, "What conditions must be satisfied in order for this principle to hold?" Einstein concluded that if there is a relative velocity between two observers, their measurements of distance and time must be different. [9]

This means that ***Reality is Relative***. For example, an object does not have a definite physical length. The length of an object depends on the velocity of the observer that is measuring it. Dimensions are relative. Time intervals are relative.

This is the essence of the Einstein theory of Relativity. In Chapter 7 we will examine the Einstein Relativity theory in detail.

Reaction to the Einstein Relativity Theory

Lorentz realized that Einstein had found the key to the enigma concerning the speed of light that was confusing the scientific world. Einstein used the same transformation equations that had been derived a year earlier by Lorentz, but interpreted them in a radically different manner. With this new interpretation, Einstein was able to make profound physical predictions that were not achieved by Lorentz.

In 1905, when Albert Einstein published his paper on Relativity, he was an unknown physicist working in the Swiss Patent Office, whereas Prof. Hendrik Lorentz was a renowned scientist. In that same year Einstein published two other landmark papers. One paper dealt with the photoelectric effect, and proved that light is quantized into small units, called photons. The other paper analyzed the statistics of the motion of gas particles, which had great application in gas experiments. .

Einstein received a Nobel Prize for his paper on the photoelectric effect. He did not receive a Nobel Prize for his much more important research on Relativity, because it was too controversial at the time.

Einstein's paper on the photoelectric effect proved that light not only acts like a wave; it also acts like a stream of particles. Einstein proved that Newton's corpuscular concept of light and Young's wave concept of light are both correct.

Chapter 7

Einstein Special Theory of Relativity

This chapter explains the basic theory of Relativity that Einstein presented in 1905. When Einstein presented his General theory of Relativity in 1916, this basic theory became known as the Einstein Special theory of Relativity.

Chapter 6 explained the fundamental principle underlying Einstein's theory of Relativity. Einstein concluded that absolute velocity does not exist. Only relative velocity has meaning. Consequently light must move at the same relative velocity with respect to all observers. Two observers moving relative to one another at a constant velocity must measure exactly the same speed of light. From this principle, Einstein derived has Special theory of Relativity.

We start our investigation of Relativity by examining the process of measuring the speed of light. To understand this we first see how the speed of sound is measured.

Measuring the Speed of Sound

Sound travels as a compression wave through the air. The speed of sound varies with the temperature of the air, and with altitude, which affects the pressure. The speed of sound at sea level and 60 °F is *340 meters per second*. To simplify our calculations, we round this off to *300 meters per second*.

We can determine the speed of sound by measuring the time for a sound wave to travel the length of a 3-meter measuring rod, as shown in Fig 7-1a. This time is equal to the length of the rod (3 meters) divided by the speed of sound (300 meters/sec), which gives 1/100 second. It is convenient to express this time in terms of milliseconds, where one

120   The Scientific Story of Creation

millisecond is one-thousandth of a second (0.001 sec). It takes 10 milliseconds (0.010 second) for the sound wave to travel the length of the 3-meter measuring rod.

In diagram (b) the measuring rod is mounted on a vehicle moving away from the sound at 50 meter/sec (112 mph). The relative speed between the sound wave and the measuring rod is now 250 meter/sec. Dividing the 3-meter rod length by this 250 meter/sec speed gives a time interval of 0.012 seconds (12 milliseconds). It takes 12 milliseconds for the sound wave to travel the length of the measuring rod.

Finally, let us assume that wind is blowing at 50 meter/sec against the sound wave, as shown in diagram (c). The measuring rod is fixed on the ground. The relative velocity between the sound wave and the measuring rod is now 250 meter/sec, just as in case (b). Therefore in case (c) it takes 12 milliseconds for the sound wave to travel the length of the measuring rod.

Sound travels with respect to the air. Consequently the time for the sound to travel the length of a measuring rod depends on the relative velocity between the rod and the air. These examples should help to explain the much more complicated effects that occur when the speed of light is measured.

Measuring the Speed of Light

Light travels about one million times faster than sound. Its speed is 300 million meters per second. This can be expressed as 300 meters per microsecond, where one microsecond is one-millionth of a second. Light takes 1/100 microsecond to travel the length of a 3-meter measuring rod. It is convenient to express this time in terms of nanoseconds. One nanosecond is one-thousandth of a microsecond, or one billionth of a second. Light takes 0.010 microsecond (or 10 nanoseconds) to travel the length of a 3-meter measuring rod.

Because light travels so fast, one must take great care in measuring the time for the light to travel the length of the measuring rod. This can be achieved as shown in Fig. 7-2. Clocks 1 and 2 are placed at the leading and trailing ends of the measuring rod. The two clocks run at exactly the same rate, and the clocks are accurately synchronized. When the light reaches the leading edge of the measuring rod, clock 1 is read. When the light reaches the trailing edge, clock 2 is read. The two clock readings are subtracted to obtain the time for the light pulse to travel the length of the measuring rod. This time difference between the two clock readings is 10 nanoseconds.

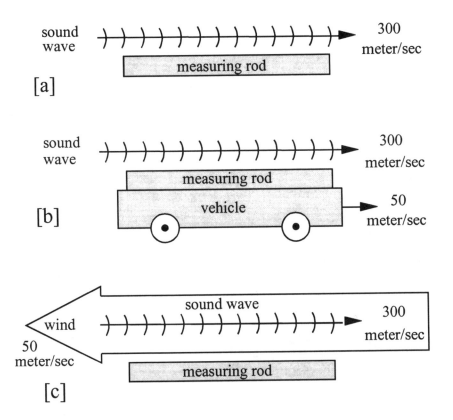

Figure 7-1: Measuring the speed of sound; [a] measuring rod fixed, no wind; [b] measuring rod moving with sound at 50 meter/sec; [c] measuring rod fixed, with wind at 50 meter/sec blowing opposite to sound propagation.

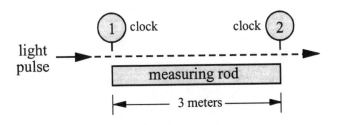

Figure 7-2: Measuring the speed of light

122   *The Scientific Story of Creation*

Synchronization of the two clocks in Fig. 7-2 could be achieved by locating the two clocks at the same point, setting their readings equal, and then moving them to the ends of the measuring rod

The Einstein Theory of Relativity

To illustrate the Einstein theory, let us apply it to a fictitious space travel episode. This example yields relativistic effects that are large enough to be conveniently discussed.

As shown in Fig 7-3, a space ship is returning from an interstellar voyage, and is now traveling toward earth at 60 percent of the speed of light. Since the speed of light is 300 meters per microsecond, the space ship velocity is 180 meters per microsecond. A light pulse is transmitted from earth, which leaves earth at a speed of 300 meters per microsecond. The earth observer A measures the speed of this light pulse in the manner that we have described. The observer B on the space ship measures the speed of the light that he receives from earth. What speed of light does the space ship observer measure?

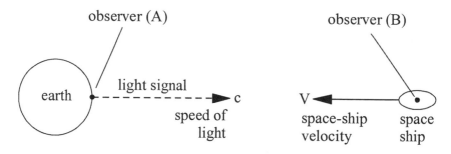

Figure 7-3: Measuring the speed of a light signal transmitted from earth to a space ship, which is returning at a velocity V equal to 60 percent of the speed of light c.

The answer is, "300 meters per microsecond", exactly the same speed of light measured by the earth observer. Both observers have identical equipment, and both find that it takes exactly 10 nanoseconds for the light to travel the length of a 3-meter measuring rod.

As was explained in Chapter 6, the constancy of the speed of light was demonstrated experimentally by the Michelson-Morley experiment. It is the fundamental postulate of the Einstein theory of Relativity.

In terms of our fanciful example, let us see how Einstein explained the constancy of the speed of light in his 1905 paper on Relativity. Einstein applied the same Lorentz transformation equations derived a year earlier by Lorentz, but interpreted them in a radically different manner.

The Principles of Relativity

In terms of Fig. 7-3, let us apply the Einstein Relativity theory to explain why the earth observer A and the observer B on the space ship measure the same value for the speed of light. Relative to the earth observer A, a measuring rod on the space ship appears to contract by the contraction factor K discussed in Chapter 6, if the measuring rod is held in the direction of the relative velocity.

Since the relative velocity V between the earth and the space ship is 60 percent of the speed of light c, the ratio V/c is 0.6 and this contraction factor K is

$$K = \sqrt{[1 - (V/c)^2]} = \sqrt{[1 - (0.6)^2]} = \sqrt{[0.64]} = 0.8$$

In accordance with Fig 6-5 of Chapter 6, this contraction factor K can be expressed geometrically as shown in Fig 7-4.

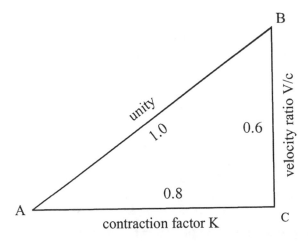

Figure 7-4: Right triangle showing in a geometric manner the relation between the V/c velocity ratio and the Special Relativity contraction factor K

The hypotenuse AB of the right triangle in Fig. 7-4 has a value of unity. The length of the side BC is equal to the velocity ratio V/c, which is 0.6. The third side AB is equal to the contraction factor K, which is 0.8. This example has been chosen to achieve a right triangle with the well-known ratios 6:8:10 (or 3:4:5).

To the earth observer A, the measuring rod on the space ship appears to contract to 80 percent of its normal length when oriented in the direction of relative velocity. To the earth observer A, the 3-meter measuring rod on the space ship appears to be 2.4 meters long, which is 80 percent of 3 meters. A dimension perpendicular to the velocity direction is not changed.

The clocks on the space ship appear to run slow when observed from earth. To the earth observer A, the clocks on the space ship appear to run at 80 percent of the rate observed by B on the space ship. A clock rate is compressed by the same contraction factor K as a spatial dimension. .

The third and final effect is that the two clocks on the space ship, which observer B considers to be synchronized, are not synchronized relative to the observer A on earth. The synchronization error in these clocks seen by observer A is

Synchronization error = $(V/c)\Delta t$ = $0.6 \Delta t$

The quantity Δt is the time for light to travel between the two clocks, as seen by observer B. Observer B finds that it takes 10 nanoseconds for light to travel the length of his measuring rod, and so Δt is 10 nanoseconds. Hence observer A sees a clock synchronization error that is 60 percent of 10 nanoseconds, which is 6 nanoseconds. This synchronization error is measured in terms of the clocks on the space ship.

Observer B on the space ship considers his two clocks to be synchronized, whereas observer A on earth considers the two clocks to be out of synchronization by 6 nanoseconds, as measured on the clocks used by observer B on the space ship.

Explanation of Constancy of the Speed of Light

Let us apply the above principles to explain why observers A and B measure the same value for the speed of light.

To observer A on earth, the light pulse emitted from earth should travel relative to the space ship at a velocity equal to (c + V), which is (300 + 180), or 480 meters per microsecond. To observer A, the space-ship measuring rod appears to be 2.4 meters long, and so the time for the light pulse to travel the length of the measuring rod is equal to the rod length (2.4 meters) divided by the speed of light relative to the space ship (480 meters per microsecond).

Relative to observer A, the light pulse should take 2.4/480 microsecond to travel the length of this measuring rod. This is 1/200 microsecond, or 5 nanoseconds. Therefore to observer A the light pulse appears to take 5 nanoseconds to travel the length of the measuring rod on the space ship.

However, the clocks on the space ship appear to run at 80 percent of the normal rate. Consequently, observer B should interpret this 5-nanosecond time interval to be 4 nanoseconds.

To the earth observer A, the two clocks used by B are out of synch by 6 nanoseconds. Adding this 6-nanosecond synchronization error to the 4-nanosecond time interval gives 10 nanoseconds. Hence, from the point of view of earth observer A, observer B on the space ship mistakenly thinks that it takes (6 + 4) nanoseconds, or 10 nanoseconds, for the light to travel the length of his measuring rod. In this manner we can explain why both observers measure the same value for the speed of light.

Symmetry of relation between observers A and B. There is no such thing as absolute velocity; there is only relative velocity. The earth observer A can assume that the earth is stationary and the space ship is moving at 60 percent of the speed of light. Likewise, the space ship observer B can assume that the space ship is stationary, and the earth is moving at 60 percent of the speed of light.

Let us assume that the space ship sends a light pulse to earth. The preceding discussion can then be applied to the space ship rather than to the earth, by considering the space ship observer to be A and the earth observer to be B. The measurements of the speed of light are symmetric.

Replacing the Observer by a Set of Coordinates

Instead of considering measurements made by an observer, we can consider measurements made relative to coordinates at the location of the observer. The principles of Relativity show how the measurements made with respect to different coordinates are related to one another.

Relativity of observation. When we compare the apparent effects experienced by the earth and space-ship observers, we should recognize that these apparent effects are real. *Reality is Relative.* There are no absolutes. The measuring rod on the space ship is 3 meters long to the space ship observer, and it is 2.4 meters long to the earth observer. Both statements are correct. The measuring rod does not have an absolute length. Nevertheless, we can still consider the rod length measured on the space ship to be special, and we call this the *proper length* of the rod.

Proper coordinates. Coordinates that move with the space ship are called *proper coordinates* for the space ship. The length of the space ship measuring rod relative to these proper coordinates is called the *proper length* of the rod. The *proper length* of the space-ship measuring rod is 3 meters.

The measuring rod on earth is 3 meters long relative to the earth observer, and 2.4 meters long relative to the space ship observer. The proper length of this measuring rod is 3 meters, which is the length that is measured from proper coordinates moving with the earth.

Four Dimensionality of Space and Time

The fact that two clocks that are synchronous relative to one observer are not synchronous relative to another observer tells us that there is no such thing as absolute time. Events that are simultaneous to one observer are generally not simultaneous to another observer moving at a different velocity.

This indicates that measurements in time and space cannot be considered separately. In Relativity, time and spatial measurements are combined together into a *four-dimensional space-time specification*. However, this does not mean that we should regard time to be a mysterious fourth dimension that is equivalent to a spatial dimension. To any observer, space and time measurements are radically different concepts.

The four dimensionality of space-time means that time and space must be considered together to obtain a precise specification. For example, a time interval between two events that is experienced by one observer can appear to be a distance interval to another observer moving at a different velocity.

Variation of Mass of an Object

Let us consider what happens when an object is accelerated until its velocity gets close to the speed of light. This effect can be achieved experimentally by accelerating an electron in an electric field, as shown by the plots in Fig. 7-5.

We assume that an electric field applies a constant force to the electron. The solid curve in Fig. 7-5 shows the resultant velocity of the electron relative to the speed of light. If the acceleration remained constant, the velocity would reach the speed of light when the relative time is 1.0.

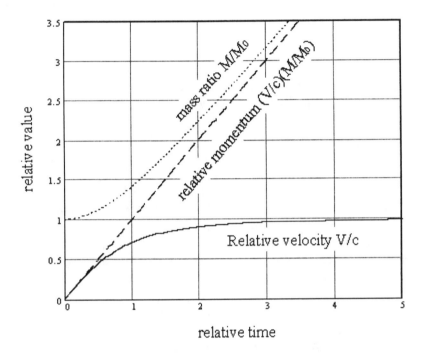

Figure 7-5: Variation with time of relative velocity, relative momentum, and relative mass of an electron accelerated by a constant force to nearly the speed of light

Initially the electron velocity increases at a constant rate. However, when the electron velocity nears the speed of light (i.e., when the V/c velocity ratio is close to unity) the slope of the velocity curve decreases, and the velocity approaches the speed of light gradually.

The slope of the velocity curve is the acceleration. According to the simple interpretation of Newton's second law, force is equal to mass times acceleration. Since the force on the electron is constant, one might expect the acceleration to be constant. However, this simple interpretation does not hold for velocities close to the speed of light.

The accurate interpretation of Newton's second law is that force is equal to the rate of change of momentum, where momentum is equal to mass M times velocity V. Since the force applied to the electron is constant, the momentum MV must increase at a constant rate. The dashed line in Fig. 7-5 shows how the momentum of the electron increases with time.

When the electron velocity is close to the speed of light, the acceleration of the electron decreases. Since the MV momentum (mass times velocity) of the electron must increase at a constant rate, the mass M of the electron must increase when the electron acceleration decreases. The dotted curve shows how the mass of the electron increases as the electron velocity nears the speed of light. The dotted curve is the ratio M/M_0, where M is the electron mass and M_0 is the mass of the electron at zero velocity, which is called the *rest mass of the electron*.

The mass of an object traveling at the velocity V is $M_0/\sqrt{[1 - (V/c)^2]}$, where M_0 is the rest mass of the object. This expression for mass would become infinite if the velocity V reached the speed of light c, and so this shows that nothing with mass can travel at the speed of light.

Proper coordinates moving with electron. Let us consider *proper coordinates* that move with the electron. Relative to these proper coordinates, the mass of the electron is its rest mass, which does not vary. If we could carry a clock in these proper coordinates, the clock would run slower as the electron velocity nears the speed of light. This effect has been observed experimentally by accelerating radioactive particles; the rate of radioactive decay decreases when the velocity of the particles gets close to the speed of light.

When the electron is very close to the speed of light, its proper clock would run very slowly. If the electron could travel at the speed of light, its proper clock would stop. This means that the electron could travel an infinite distance (as measured externally) in zero *proper time* (as measured on a clock traveling with the electron).

Converting Matter into Energy

The increase in mass of an object as it approaches the speed of light indicated to Einstein that adding energy to an object increases its mass. This in turn showed that mass M can be converted into energy E in accordance with the following famous Einstein formula:

$$E = Mc^2$$

This formula explained the source of the energy radiated by the sun, and eventually led to the atomic and hydrogen bombs. The equation shows that one gram of mass M is equivalent to 25 million kilowatt-hours of energy E. This means that 25 million kilowatt hours of energy are released for every gram of matter that is converted into energy.

The sun achieves its energy by fusing four hydrogen atoms to form one helium atom. The atomic weight of helium is 3.971 times that of hydrogen, and so the ratio of helium mass to hydrogen mass is 3.971/4, which is 0.99275. The helium atom has 99.275 percent of the mass of the four hydrogen atoms that form it. The remaining 0.725 percent of the hydrogen mass is converted into energy. Taking 0.725 percent of 25 million kilowatt-hours gives 181,250 kilowatt-hours. This shows that the conversion of one gram of hydrogen into helium releases 181,250 kilowatt-hours of energy. One gram is 1/3 of the weight of a United States penny.

It is remarkable that the abstract principles of Special Relativity allowed Einstein to explain the source of the enormous energy radiated by our sun. *The fact that matter can be converted into energy, in accordance with this Einstein formula, demonstrates that Einstein's Relativity principles are correct. It proves that Reality is Relative.*

The Principle of Covariance

The Relativity principles that we have applied to relate the measurements of the observer in the space ship to the observer on earth are expressed in rigorous mathematical form by a set of formulas known as the Lorentz transformation equations. These equations were first derived in 1904 by Hendrik Lorentz, and so bear his name. However Lorentz interpreted his equations in terms of motion relative to the hypothetical aether. It was Einstein who established the basic principles of Relativity.

Let us consider the application of the Lorentz transformation equations to a physical problem. As was stated earlier, we can imagine that each observer in our discussion is replaced by a set of coordinates. Instead of considering measurements made by particular observers, we can consider measurements made relative to coordinate systems at the locations of the observers. The Lorentz transformation equations allow one to translate measurements made relative to one coordinate system to another, in such a manner that physical consistency is maintained.

We have been investigating the resolution of an inconsistency in measuring the speed of light. However, the Lorentz transformation equations have more general applicability. Maxwell's electromagnetic field equations characterize the physical laws of electricity and magnetism. Lorentz proved that Maxwell's equations are invariant when modified by his Lorentz transformation equations. What this means is that the two observers, moving at different velocities, experience exactly the same electrical and magnetic laws when they make measurements relative to their separate coordinate systems.

Thus the Lorentz transformation equations allow one to translate physical measurements from one coordinate system to another in such a manner that the same physical laws are experienced by observers in the two coordinate systems.

This concept can be stated in a general manner as the *Principle of Covariance*. This principle states that the laws of physics should be expressed so that that they are independent of the coordinate system. The fundamental goal of Einstein's research in developing his General theory of Relativity was to achieve *Covariance* in his transformation equations.

Special Relativity deals with measurements made by observers in separate coordinate systems moving at different but constant velocities. Under this condition, the speed of light is the same in both coordinates. However when the velocity changes, so that acceleration occurs, Special Relativity no longer applies exactly, and the speed of light can be different in the two coordinate systems. Einstein showed that acceleration and gravity are equivalent. Therefore the Lorentz transformation equations of Special Relativity do not apply exactly in a gravitational field.

Although the Lorentz transformation equations of Special Relativity do not apply exactly in a gravitational field or under acceleration, they still give extremely close approximations for almost all applications within our solar system. It is only in very restricted cases that these theoretical limitations are important.

Nevertheless the theoretical limitations of Special Relativity are fundamental, and Einstein was compelled to eliminate them. In 1916, after 11 years of intense research, he achieved this with his General theory of Relativity. His General theory achieves *Covariance* of physical laws for coordinates that are accelerating or are in gravitational fields.

Chapter 8

The Einstein Theory of Gravity

Generalizing the Relativity Principle

Einstein derived several profound conclusions from his basic Relativity concept, but he soon found that it had a serious weakness. The theory applies exactly only when the velocities of the two observers are constant. When acceleration occurs, which means that the velocity is changing, Einstein found that the speed of light is not exactly constant. He also concluded that acceleration and gravity are equivalent, and so a gravitational field changes the speed of light. Since constancy of the speed of light is the unifying principle for his basic theory of Relativity, Einstein needed a new principle to generalize the Relativity concept. He found this in the complex mathematics of *tensor analysis* developed by Gregorio Ricci, which was based on a geometric principle derived by Bernhard Riemann.

In 1852, the German mathematician, Bernhard Riemann (1826-1866), presented a basic mathematical principle for characterizing curved space. Riemann derived the formulas for the *geodesic* and the *metric equation*, which describe the shortest distance between two points in curved space. Riemann was unable to develop these geometric principles in detail, because he contracted tuberculosis in 1862, and died four years later at age 39.

The Italian mathematician, Gregorio Ricci (1853-1925), used the curved-space principle of Riemann as the foundation for a comprehensive mathematical theory, which Ricci called the *absolute differential calculus*, and is now called *tensor analysis*. *Tensor analysis* is a mathematical theory that generalizes calculus to apply to curved space. Ricci [11] published his mathematical theory in 1901 with the help of his student, Tullio Levi-Civita (1873-1941). In 1923, Levi-Civita published an updated version of this theory. An English translation of this is available as a Dover reprint. [5]

Tensor analysis has a unique formula for translating a tensor from one coordinate system to another. Therefore, by expressing Relativity principles in terms of tensors, Einstein had the basis for achieving *covariance*, whereby the same laws of physics hold in all coordinate systems, regardless of velocity and acceleration, and regardless of gravitational fields. However, tensor analysis is very complicated. It took 11 years of intensive research before Einstein [10] was able to publish his *General theory of Relativity* in 1916. His basic Relativity theory presented in 1905 was then called the *Special theory of Relativity*.

Einstein concluded that the concept of gravitational force presented by Newton is incompatible with Relativity, because it represents a force operating instantaneously at a distance; whereas relativistic effects propagate at the speed of light. Instead, Einstein described gravity as a curvature of space. Matter causes space to be curved, and this space curvature produces the effect that we interpret as gravitational force. The Riemannian principle for specifying curved space, which is incorporated in tensor analysis, provided the basis for characterizing gravity in General Relativity.

In 1916 Einstein presented his General theory of Relativity, which was specified in terms of a tensor formula called the *Einstein gravitational field equation*. This tensor formula represents 10 independent equations. Because of the great complexity of this formula, Einstein was only able to derive approximate solutions from it. However, Karl Schwartzschild (1873-1916), who was cooperating with Einstein, applied the Einstein formula to a simple physical model of a star, and thereby achieved an exact solution. Einstein also published the Schwartzschild solution in 1916. Unfortunately Karl Schwartzschild died suddenly from disease even before his famous analysis was printed. He was a German army officer at the Russian front during World War I.

General Relativity theory incorporates the Special Relativity effects caused by velocity and the General Relativity effects caused by gravity and acceleration. Within the weak gravitational fields of our solar system, the relativistic effects due to gravity are very small, but are measurable. Based on the Schwartzschild solution, Einstein devised the following three tests to verify his General theory of Relativity:

(1) When a light ray passes close to the sun, it should be deflected by 1.8 arc seconds.

8. Einstein Theory of Gravity   133

(2) A gravitational field causes a clock to run slower, and therefore causes the excited elements on the surface of the sun to oscillate at lower frequency, thereby generating spectra of longer wavelength. The gravitational field of our sun should cause the spectra of light from the sun surface to shift toward the red end of the spectrum by 2.1 parts per million of wavelength (i.e., by 2.1×10^{-6}).

(3) The planet Mercury has a highly elliptical orbit. The axis of the Mercury orbit advances (or rotates) by 1.39 arc seconds per orbit. Of this advance of the orbit axis, 1.29 arc seconds can be explained with Newton's laws by considering the gravitational attraction of other planets. A residual error of 0.10 arc second per orbit remained, which was explained by the Einstein General theory of Relativity.

These three tests were implemented, and the results established the validity of the Einstein General theory of Relativity.

These measurable effects of General Relativity are very small: an advance of only 0.10 arc second per orbit of Mercury; a 1.8 arc second deflection of a light beam passing close to the sun, and a gravitational redshift of only 2.1 parts per million in light emitted from the sun. Hence one might wonder why Einstein worked so hard to achieve his theory, and why General Relativity is so highly regarded. *The answer is that this generalization was essential to provide a solid theoretical foundation for the Relativity principle that is embodied in Special Relativity.*

When the predictions of General Relativity were verified, Einstein achieved great fame. After that time, Einstein did little with his General Relativity theory. Special Relativity is very much easier to apply, and has wide applicability. During Einstein's lifetime, General Relativity, with its very complicated tensor analysis, served primarily as a theoretical foundation for justifying Special Relativity.

Einstein could apply General Relativity only to very simple cases. With more complicated applications, the equations of General Relativity can yield millions of terms, and so cannot be solved analytically. In the 1960's, a decade after Einstein's death, powerful computers became available, which could be used to apply the Einstein General Relativity theory to complicated physical models. Since then, many hundreds of mathematicians, physicists, and engineers in academic positions have devoted their careers to computer solutions of the Einstein gravitational field equation. This effort has been devoted to Big Bang studies of cosmology, because that is the only area that can use this expertise.

Applying the Equivalence Principle

Einstein concluded that the effects produced by gravity and acceleration are equivalent. To generalize his Relativity theory, he applied the principles of Relativity under conditions of acceleration, and related these results to a gravitational field. Let us see how he did this.

Equivalence between Acceleration and Gravity

As shown in Fig. 8-1, Einstein considered two identical elevators. One in diagram (b) rests on the ground, and the other in (a) is located in space, where gravitational force is negligible. Einstein postulated that the elevator in space is being pulled upward with the acceleration of gravity g that is experienced on earth. Today we assume more realistically that a rocket motor under elevator (a) is pushing the elevator upward with the acceleration g.

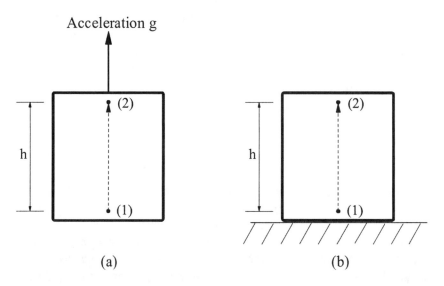

Figure 8-1: Elevator (a) is accelerating in free space; elevator (b) is fixed on earth. Light travels from point (1) to point (2).

The meaning of the acceleration of gravity g was explained in Chapter 5. When an object is allowed to fall on earth, it drops with a constant acceleration of gravity g of about 10 meter/sec per sec. The velocity of a falling object is proportional to time, provided that the force exerted by air is negligible. When an object is not allowed to fall, it exerts a force on the floor, which we call weight. The weight W of an object is equal to its mass M multiplied by the acceleration of gravity g.

Since the elevator in diagram (a) is pushed upward by a rocket at the same acceleration of gravity g experienced on earth, an object within elevator (a) would exert the same weight force on the floor that it would exert if it were located within the fixed elevator (b) located on earth. The conditions inside elevators (a) and (b) are the same. If the elevators are closed, a scientist could not tell from experiments performed within the two elevators whether he is in elevator (a) or elevator (b). This is the *Principle of Equivalence* that was proposed by Einstein, which allowed him to relate the effects of acceleration and gravity.

Redshift Produced by Gravity

Suppose that a light pulse is emitted from the floor of elevator (a) at point (1) and is received at the ceiling at point (2). During the propagation time of the light, the velocity of the elevator increases, because the elevator is accelerating. Therefore the upward velocity of the receiver at point (2) is greater than that of the emitter at point (1), as far as the light pulse is concerned. Point (2) appears to be moving away from point (1), and so the light is redshifted as it moves from (1) to (2). The wavelength of the light received at (2) is greater than the wavelength emitted at (1).

Table 8-1 shows the steps for calculating this change of wavelength. The calculations of this table are approximate. Item (a) shows the propagation time Δt for light to travel from (1) to (2). The propagation time Δt is approximately equal to (h/c), where h is the height of point (2) above point (1) and c is the speed of light.

Table 8-1: *Calculation of approximate redshift caused by an accelerating elevator or by a fixed elevator in a gravitational field.*

(a) Propagation time of light (Δt)	h/c
(b) Velocity change during light propagation (ΔV)	$g\Delta t = gh/c$
(c) Velocity ratio ($\Delta V/c$)	gh/c^2
(d) Doppler redshift of light ($\Delta\lambda/\lambda$)	$\Delta V/c = gh/c^2$
(e) Wavelength ratio: $\lambda_2/\lambda_1 = 1 + (\Delta\lambda/\lambda)$	$1 + (gh/c^2)$

During the light propagation time Δt, the velocity increases by an amount ΔV equal to the acceleration g multiplied by the time interval Δt. Hence this velocity change ΔV is equal to $g\Delta t$, which is equal to gh/c, as shown in item (b). The velocity change ΔV is divided by the speed of light c to obtain the velocity ratio $\Delta V/c$, which is equal to gh/c^2, as shown in item (c).

The velocity change ΔV produces a Doppler wavelength shift $\Delta\lambda$. The wavelength ratio $\Delta\lambda/\lambda$, which is called the *redshift*, is the ratio of the wavelength shift $\Delta\lambda$ to the normal wavelength λ. The $\Delta\lambda/\lambda$ redshift is approximately equal to the velocity ratio $\Delta V/c$, and so is approximately equal to gh/c^2, as shown in item (d).

Since the conditions in elevators (a) and (b) are equivalent, a light pulse emitted at point (1) in elevator (b) experiences a redshift when it is received at point (2). The gravitational field in elevator (b) affects the wavelength of light in exactly the same manner as the acceleration applied to elevator (a). Therefore, in accordance with item (d) of Table 8-1, the gravitational field in elevator (b) produces a redshift $\Delta\lambda/\lambda$ approximately equal to gh/c^2 as the light pulse propagates from point (1) to point (2).

The wavelength received at point (2) is denoted λ_2 and the wavelength emitted at point (1) is denoted λ_1. The wavelength λ_2 is equal to $(\lambda_1 + \Delta\lambda)$, where $\Delta\lambda$ is the wavelength shift. Hence the wavelength ratio (λ_2/λ_1) is equal to $(1 + \Delta\lambda/\lambda_1)$, which can be approximated as $(1 + \Delta\lambda/\lambda)$. Since $\Delta\lambda/\lambda$ is approximately equal to gh/c^2, λ_2/λ_1 is approximately equal to $[1 + (gh/c^2)]$, as shown in item (e).

The calculations shown in Table 8-1 were implemented by Einstein, and were an important step in the development of General Relativity. These calculations are approximate. Yilmaz examined this analysis and realized that he could implement it exactly. This exact analysis resulted in the Yilmaz theory of gravity and is given in Appendix E.

Effect of Gravity on Clock Rate

Let us place a clock at point (1) in elevator (a), and time the emitted light pulses with the ticking of the clock, so that one pulse is emitted every nanosecond. The spacing between the light pulses received at (2) will be greater than one nanosecond. To understand this, imagine that you are running away from a train of light pulses emitted by a fixed source. As you run, the distance between you and the light source increases, and so the number of light pulses along the transmission path increases. Consequently, the rate at which you receive light pulses must decrease.

This reasoning shows that the acceleration of the elevator (a) causes a clock at (1) to tick more slowly when observed at (2). Similarly the gravitational field in elevator (b) causes a clock on the floor of the elevator at (1) to tick more slowly when observed from point (2) at the ceiling. When observed from the ceiling at (2), the clock at (1) on the floor appears to be slowed approximately by the factor $[1 - (gh/c^2)]$.

Now, let us assume that a light pulse is emitted at the ceiling of the elevator and is observed at the floor. In this case, point (1) appears to be moving toward (2). A Doppler wavelength shift is observed, but the spectrum is now shifted toward the blue end of the spectrum (toward shorter wavelength). Hence when light travels from the ceiling to the floor, the spectrum experiences a $\Delta\lambda/\lambda$ blueshift approximately equal to (gh/c^2). Similarly, when a clock located on the elevator ceiling is observed at the floor, the clock appears to run fast. The clock rate observed at the floor is approximately equal to the clock rate at the ceiling multiplied by $[1 + (gh/c^2)]$.

The gravitational field of the sun causes a redshift in the sun spectrum, when observed from earth. On earth we are located at the ceiling of the elevator (further from the gravitational mass of the sun) and the light source is located at the floor of the elevator (on the surface of the sun). Also a clock on the sun surface would appear to run more slowly when observed from earth.

If we were stationed on the surface of the sun (which is difficult to achieve!) and observed a light emitted from earth, the spectrum of the light would be blue-shifted. Similarly, a clock on earth would appear to run faster when observed from the surface of the sun.

This reasoning is similar to the analysis that Einstein made to determine the changes of wavelength and clock rate caused by a gravitational field. These calculations by Einstein were approximate.

Effect of gravity on dimensions and speed of light. Einstein performed other approximate analyses to prove that acceleration and gravity cause a dimension to contract and the speed of light to decrease. He needed to integrate into a rigorous theory all of these effects, along with the Special Relativity effects produced by velocity. To achieve this result, he applied the mathematical theory of *tensor analysis* that had been developed by Ricci and Riemann.

Application of Tensor Analysis to General Relativity

To investigate the Einstein theory, we must become familiar with the tensor. The mathematical meaning of a tensor is explained in Appendix D, but that information is not necessary. The reader should understand the tensor symbolism and the physical significance of the various tensors used in the Einstein theory. These tensors are:

metric tensor (g_{ab}, g^{ab})
Ricci tensor $(R_{ab}, R_a{}^b)$
Einstein tensor $(G_a{}^b)$
energy-momentum tensor $(T_a{}^b, T^{ab})$

The expressions in parentheses give the mathematical tensor forms that are generally used. These forms will be explained momentarily.

The *metric tensor* specifies the shortest distance between two points in curved space. The *Ricci tensor* and the *Einstein tensor* are closely related tensors that specify the curvature of space. The *energy-momentum tensor* specifies the characteristics of matter and energy.

The Metric Tensor

The starting point of tensor analysis is the metric tensor, which specifies the shortest distance between two points in curved space.

The metric tensor is denoted g_{ab}. The subscripts a, b are called *indices*, where each *index* can represent 0, 1, 2, or 3. Since index (a) has four possible values (0, 1, 2, 3) and index (b) has these same possible values, the metric tensor has 4x4 or 16 elements. These elements are arranged as follows in an array called a "matrix":

$$\begin{vmatrix} g_{00} & g_{01} & g_{02} & g_{03} \\ g_{10} & g_{11} & g_{12} & g_{13} \\ g_{20} & g_{21} & g_{22} & g_{23} \\ g_{30} & g_{31} & g_{32} & g_{33} \end{vmatrix}$$

The index 0 specifies the time dimension, and the indices 1, 2, 3 specify the three spatial dimensions. The metric tensor g_{ab} has the symmetry property that g_{ba} is equal to g_{ab}. For example g_{21} is equal to g_{12}. Because of this symmetry, 6 of the 16 elements of the metric tensor are redundant. ***Therefore, in its general form, the metric tensor g_{ab} has 10 independent elements.***

The 4 elements along the diagonal of the matrix, from the upper left to the lower right, are called the "diagonal elements". These 4 diagonal elements are $g_{00}, g_{11}, g_{22}, g_{33}$. The other 12 elements are called the non-diagonal elements of the tensor. A tensor in which the non-diagonal elements are all zero is called a "diagonal tensor".

If the metric tensor is diagonal, it is possible to develop analytical solutions of the Einstein theory. However, if the metric tensor is not diagonal, the equations for the Einstein theory can have millions of terms, and so cannot be solved analytically; a computer is required. ***Since computers were not available during Einstein's lifetime, Einstein had to limit the application of his General Relativity theory to very simple physical models that could yield diagonal metric tensors.***

A diagonal metric tensor has the following form

$$\begin{vmatrix} g_{00} & 0 & 0 & 0 \\ 0 & g_{11} & 0 & 0 \\ 0 & 0 & g_{22} & 0 \\ 0 & 0 & 0 & g_{33} \end{vmatrix}$$

The element g_{00} describes the time-component of the metric tensor and the other elements describe the three-dimensional spatial components.

There are three different forms of a tensor, which are indicated by the indices. For the metric tensor these forms are:

covariant tensor	g_{ab}	subscripted indices
contravariant tensor	g^{ab}	superscripted indices
mixed tensor	$g_a{}^b$	subscripted and superscripted indices

The mixed form of the metric tensor $g_a{}^b$ is seldom used, but this form is commonly used for other tensors. There are complicated theoretical issues associated with *covariant, contravariant,* and *mixed* tensors, but these need not concern us. We merely need to understand that there are three tensor forms, and that there are precise formulas for converting from one tensor form to another.

If the covariant metric tensor g_{ab} is diagonal, the contravariant metric tensor g^{ab} is also diagonal, and the four diagonal elements of g^{ab} are the reciprocals of the corresponding elements of g_{ab}. Thus if the covariant metric tensor g_{ab} is diagonal, the nondiagonal elements of the contravariant metric tensor g^{ab} are all zero, and the diagonal elements are

$$g^{00} = 1/g_{00} \; ; \; g^{11} = 1/g_{11} \; ; \; g^{22} = 1/g_{22} \; ; \; g^{33} = 1/g_{33}$$

Converting Form of a Tensor

If the elements of the covariant and contravariant metric tensors (g_{ab} and g^{ab}) are known, one can readily convert any other tensor from one of its forms (covariant, contravariant, or mixed) to another form by means of simple equations.

The Ricci and Einstein Curvature Tensors

In the Einstein General theory of Relativity, a gravitational field is defined as a curvature of space. The curvature of space is specified by two closely related tensors, the Ricci tensor R_{ab} and the Einstein tensor G_{ab}. If space has no gravitational field, and hence no curvature, all elements of the Ricci and Einstein curvature tensors are zero.

If the covariant and contravariant metric tensors (g_{ab} and g^{ab}) are known, one can calculate the Ricci tensor R_{ab} from precise but *very complicated* formulas. *If the metric tensor is diagonal, the Ricci tensor can be calculated analytically, although it may take a skilled mathematician a few weeks to solve the equations without error. On the other hand, if the metric tensor is not diagonal, the equations can result in millions of terms and so can only be solved on a computer.*

After the covariant Ricci tensor R_{ab} is obtained, this is converted to the mixed form $R_a{}^b$. If the covariant and contravariant metric tensors (g_{ab} and g^{ab}) are known, this step for calculating $R_a{}^b$ is easy.

The next step is to compute the variable R, which is the sum of the four diagonal elements of the mixed Ricci tensor $R_a{}^b$. The variable R is

$$R = R_0{}^0 + R_1{}^1 + R_2{}^2 + R_3{}^3$$

The nondiagonal elements of the mixed Einstein tensor $G_a{}^b$ are the same as the corresponding elements of the mixed Ricci tensor $R_a{}^b$. The diagonal elements of the Einstein tensor $G_a{}^b$ are obtained by subtracting ½R from the corresponding elements of the Ricci tensor $R_a{}^b$. Hence the Einstein tensor is computed from the Ricci tensor as follows:

Nondiagonal elements: $\quad G_a{}^b = R_a{}^b$

Diagonal elements: $\quad G_a{}^b = R_a{}^b - \tfrac{1}{2} R$

The Energy-Momentum Tensor

The energy-momentum tensor describes the characteristics of matter and energy. Its common forms are the mixed and contravariant tensors $T_a{}^b$ and T^{ab}. *The contravariant energy-momentum tensor T^{ab} is computed from the physical characteristics of matter and energy. There are precise rules for performing this calculation.*

The Einstein Gravitational Field Equation

Einstein specified his General theory of Relativity by his gravitational field equation, which has the following tensor formula:

$$G_a{}^b = -8\pi T_a{}^b$$

In this formula the indices a, b can take on the four values (0, 1, 2, 3). Hence this tensor formula represents 16 individual equations. For example, $G_0{}^1$ is equal to $-8\pi T_0{}^1$, $G_2{}^3$ is equal to $-8\pi T_2{}^3$, etc. Because of tensor symmetry the expression for $G_2{}^3$ can be derived from $G_3{}^2$, etc., and so 6 of the 16 equations are redundant. *Thus the Einstein gravitational field equation is a tensor formula that represents 10 independent equations.*

Outline of Einstein Theory Calculations

Table 8-2 outlines the steps that are involved in calculating the contravariant energy-momentum tensor T^{ab} from the covariant metric tensor g_{ab}. This table shows that this calculation involves great mathematical complexity, but there is nothing mysterious about the process. The steps are precisely specified by mathematical formulas.

In step (1) of Table 8-2, the contravariant metric tensor g^{ab} is calculated from the covariant metric tensor g_{ab}. If the covariant metric tensor is diagonal, this computation is simple. Otherwise it is complicated.

In step (2) the covariant Ricci tensor R_{ab} is calculated from the covariant and contravariant metric tensors (g_{ab}, g^{ab}). As was stated earlier, this computation is very complicated, even when the metric tensor is diagonal. If the metric tensor is not diagonal, this computation results in millions of terms and so can only be performed with a computer.

Table 8-2: *Steps for calculating the energy-momentum tensor from the metric tensor*

From	Calculate	Tensor Name
(1) g_{ab}	g^{ab}	contravariant metric tensor
(2) g_{ab}, g^{ab}	R_{ab}	covariant Ricci tensor
(3) g_{ab}, g^{ab}, R_{ab}	$R_a{}^b$	mixed Ricci tensor
(4) $R_a{}^b$	$G_a{}^b$	mixed Einstein tensor
(5) $G_a{}^b$	$T_a{}^b$	mixed energy-momentum tensor
(6) $g_{ab}, g^{ab}, T_a{}^b$	T^{ab}	contravariant energy-momentum tensor

In step (3) the covariant Ricci tensor R_{ab}, along with the metric tensors g_{ab}, g^{ab}, are used to calculate the mixed form of the Ricci tensor $R_a{}^b$. This is a simple calculation. If the metric tensors g_{ab}, g^{ab} are known, any tensor can be readily converted from one of its forms (covariant, contravariant, or mixed) to another.

In step (4) the mixed Einstein tensor $G_a{}^b$ is calculated from the mixed Ricci tensor $R_a{}^b$. As was shown, this is a simple calculation. The Einstein and Ricci tensors both characterize the curvature of space and are closely related to one another.

Step (5) applies the Einstein gravitational field equation to calculate the elements of the mixed energy-momentum tensor $T_a{}^b$ from those of the mixed Einstein tensor $G_a{}^b$.

To relate the energy momentum tensor to the physical model, one needs the contravariant energy-momentum tensor T^{ab}. In step (6) the mixed energy-momentum tensor $T_a{}^b$ is combined with the metric tensors g_{ab}, g^{ab} to calculate the contravariant form of the energy-momentum tensor T^{ab}.

Backward calculation. Table 8-2 outlines the steps for calculating the contravariant energy-momentum tensor T^{ab} from the covariant metric tensor g_{ab}. However, when the Einstein theory is applied, the metric tensor is not known. The energy-momentum tensor T^{ab} is derived from the characteristics of the physical model. *One must solve the complicated equations outlined in Table 8-2 backward. One must calculate the elements of the metric tensor g_{ab} that yield the known elements of the contravariant energy-momentum tensor T^{ab}.*

If the metric tensor is not diagonal, the Ricci and Einstein tensors can have millions of terms. Therefore, analytical solutions are possible only for very simple physical models that yield diagonal metric tensors.

Prior to the presentation of his General theory of Relativity in 1916, Einstein was only able to achieve approximate solutions to his equations. Karl Schwartzschild, who was cooperating with Einstein, derived an exact solution for a very simple physical model of a star. Let us examine the Schwartzschild solution.

The Schwartzschild Solution

Physical model of a star. To obtain his energy momentum tensor, Schwartzschild considered a simple physical model of a star. He assumed that the star has no viscosity and has a constant density of matter. Since a star is gaseous, the assumption of constant mass density is highly inaccurate, but was necessary to achieve equations that could be solved analytically.

The surface of our sun has about the same density as the earth's atmosphere, which is 1/1000 of the density of water. The center of the sun has 100 times the density of water. Hence the density of the sun varies with radius by a factor of 100,000. Nevertheless, the analysis by Schwartzschild approximated the sun as a fluid having a constant density of matter. This very unrealistic approximation was essential. Without this approximation, the Einstein equations could not be solved analytically.

Calculation of the Einstein tensor $G_a{}^b$. From this very simple physical model of a star, Schwartzschild derived the corresponding contravariant energy-momentum tensor T^{ab}. This tensor is diagonal, which means that the 12 nondiagonal elements are all zero.

The contravariant form T^{ab} must be converted to the mixed form $T_a{}^b$. This conversion is simple if the metric tensor g_{ab} is known, but at this point the metric tensor was not known. With a brilliant calculation, Schwartzschild was able to derive the mixed energy-momentum tensor $T_a{}^b$ from the contravariant form T^{ab} without specifically knowing the values of the metric tensor g_{ab}.

The values for the four diagonal elements of the mixed energy-momentum tensor $T_a{}^b$ are shown in Table 8-3. The other (nondiagonal) elements of this tensor are all zero.

The quantity p in Table 8-3 represents pressure inside the star, which varies with radius. The quantity ρ is the density, which was specified to be constant. Parameter G is the gravitational constant in Newton's law of gravity, and c is the speed of light. (Do not confuse the gravitational constant G with the Einstein tensor $G_a{}^b$.)

Table 8-3: *Non-zero elements of the energy-momentum tensor $T_a{}^b$ and the Einstein tensor $G_a{}^b$ for the interior Schwartzschild solution.*

element	value	element	value
$T_0{}^0$	$\rho(G/c^2)$	$G_0{}^0$	$-8\pi\rho(G/c^2)$
$T_1{}^1, T_2{}^2, T_3{}^3$	$-p(G/c^2)$	$G_1{}^1, G_2{}^2, G_3{}^3$	$8\pi p(G/c^2)$

By applying the Einstein gravitational field equation to these elements of the energy-momentum tensor, the elements of the Einstein tensor $G_a{}^b$ were calculated, which are also shown in Table 8-3. This table gives the diagonal elements of the energy momentum tensor and the Einstein tensor. The nondiagonal elements of both tensors are zero.

The Schwartzschild analysis consists of two solutions. The *interior solution* applies inside the star, and the *exterior solution* applies outside the star. The values in Table 8-3 are for the *interior solution*, which holds inside the star. In the vacuum of space outside the star, the energy momentum tensor is zero, because there is no matter outside the star.

Therefore in the exterior Schwartzschild solution, all elements of the energy-momentum tensor $T_a{}^b$ and the Einstein tensor $G_a{}^b$ are zero.

Calculating the metric tensor g_{ab}. Schwartzschild had to find the g_{ab} metric tensor elements that would yield his calculated values for the elements of the Einstein tensor $G_a{}^b$. He did this by assuming general formulas for the metric tensor elements having unknown parameters.

From these general formulas of the g_{ab} metric tensor elements, Schwartzschild calculated the corresponding expressions for the elements of the Einstein tensor $G_a{}^b$. He set these expressions equal to the values he had calculated for his *interior* and *exterior* solutions. He solved the resultant equations to obtain the values of the unknown parameters in the formulas for the metric tensor elements.

In this manner, Schwartzschild developed his famous solution for Einstein's General theory of Relativity This solution gives the elements of the metric tensor g_{ab} that characterize a very simple physical model of a star. As will be shown in Chapter 10, from these g_{ab} metric tensor elements one can calculate the relativistic effects that are produced by the gravitational field of the star. For example, the relative clock rate is equal to $\sqrt{[g_{00}]}$.

8. Einstein Theory of Gravity   145

This analysis by Schwartzschild was a brilliant achievement. Einstein could only derive approximate solutions of his theory. If Karl Schwartzschild had lived he would probably have been one of the giants of 20th-century physics. Details of the Schwartzschild analysis are given in *Addendum* [3], Chapter 2 and Appendix C.

Computer Solutions of the Einstein Theory

The Einstein General theory of Relativity can be solved analytically only when the metric tensor is diagonal. If the metric tensor is not diagonal, the calculations yield millions of terms. During Einstein's lifetime, the theory could be applied only to very simple physical models that resulted in diagonal metric tensors.

In the 1960's, computers became available that could solve the Einstein gravitational field equation for a complex physical model. However, even with a powerful computer the solution of the Einstein gravitational field equation for a general model is very difficult, because the equations can yield millions of terms and must be solved backward. A complicated iterative computer program is needed to achieve this backward calculation.

The iterative program starts with approximate values for the metric tensor elements, and from these it computes the corresponding elements of the Einstein tensor. The computed elements of the Einstein tensor are compared with the desired elements, and the differences are used to change the metric tensor elements in such a way as to minimize the differences. The program cycles through this process until the computed and desired elements of the Einstein tensor match.

The iterative program may implement billions of cycles before a solution is obtained. Sophisticated mathematical algorithms have been developed to achieve iterative computer programs that converge to solutions. Since the 1960's, many hundreds of scientists in academic positions have devoted their careers to the task of solving the formidable Einstein equations on the computer. Highly advanced analytical procedures have been devised to achieve these computer solutions.

Albert Einstein displayed a profound genius in developing his General theory of Relativity, but a similar genius is not needed to apply his theory. The mathematical formulas associated with the calculations of the Einstein theory are precisely specified. Although it takes great mathematical skill to solve these equations on a computer, these computational skills do not place a theoretician solving the Einstein equations in the same intellectual category as Albert Einstein.

Chapter 9

The Yilmaz Theory of Gravity

Derivation of the Yilmaz Solution

The Elements of the Metric Tensor

As was shown in Chapter 8, Einstein proved that when light rises against a gravitational field it experiences a redshift $\Delta\lambda/\lambda$ approximately equal to gh/c^2, where g is the acceleration of gravity and h is the height through which the light rises. This shows that the ratio λ_2/λ_1 of the wavelength λ_2 at the top, to the wavelength λ_1 at the bottom, is approximately equal to $[1 + (gh/c^2)]$. This approximate result derived by Einstein is shown as item (1) in Table 9-1.

Huseyin Yilmaz examined this calculation by Einstein as part of his PhD thesis research at the Massachusetts Institute of Technology in the early 1950's. He found that he could perform the analysis exactly. He derived an exact formula for this λ_2/λ_1 wavelength ratio, which is shown in item (2) of Table 9-1. This analysis is given in Appendix E.

Table 9-1: *Comparison of wavelength ratios produced by gravitational field calculated approximately by Einstein and exactly by Yilmaz*

Source of solution	λ_2/λ_1
(1) Approximate Einstein solution	$[1 + (gh/c^2)]$
(2) Exact Yilmaz solution	$\exp[gh/c^2]$

The formula in Table 9-1 derived by Yilmaz is undoubtedly strange to many readers, but is a well-recognized function to those familiar with calculus. This function is called the *exponential of $[gh/c^2]$*. Since we will observe the effect of this function in graphical plots, the reader does not need to understand its meaning.

The Yilmaz expression of item (2) approximates the Einstein expression of item (1) if the term (gh/c^2) is much less than unity. Therefore the Yilmaz analysis agrees with Einstein's approximation.

On the other hand, the Yilmaz analysis is exact and therefore provides an exact solution to the relativistic principles that Einstein established in developing his theory. From the λ_2/λ_1 wavelength ratio in Table 9-1, Yilmaz was able to derive an exact expression for the metric tensor element g_{00}, which describes the relativistic effects associated with time. This derivation is shown in Appendix E.

To obtain the other metric tensor elements, Yilmaz postulated that the speed of light measured locally is independent of direction. He proved that this postulate requires that the following conditions hold in rectangular coordinates:

$$g_{11} = -1/g_{00}$$

$$g_{33} = g_{22} = g_{11}$$

The derivation to prove these relations is given in the *Addendum* [3] Section 3.5. With these formulas, Yilmaz was able to calculate all of the metric tensor elements, and thereby achieved a complete and exact solution to the principles of Relativity.

Yilmaz mailed his results to Einstein. However, Albert Einstein was too sick to read them, and died soon thereafter.

The Yilmaz Gravitational Field Equation

With the elements of the metric tensor known, Yilmaz could calculate rigorously the corresponding elements of the Einstein tensor $G_a{}^b$. This yielded the gravitational field equation for the Yilmaz theory, which has the form

Yilmaz: $\qquad G_a{}^b = -8\pi T_a{}^b - 2t_a{}^b$

This differs from the gravitational field equation of the Einstein theory in that it has the additional term $-2t_a{}^b$, where $t_a{}^b$ is called the *stress-energy tensor for the gravitational field*.

As will be explained in Chapter 13, the fact that the Einstein theory lacks a stress-energy tensor for the gravitational field is a severe weakness. Appendix E shows that Einstein searched for such a tensor, but could only isolate a *pseudo-tensor*. Einstein believed that his pseudo-tensor describes the *"energy components of the gravitational field"*. Nevertheless, it is not a true tensor, and so could not be used in the Einstein gravitational field equation.

The gravitational field equations of both the Einstein and Yilmaz theories have the *energy-momentum tensor* T_a^b. However, to simplify the equation that specifies his theory, Yilmaz replaces $4\pi T_a^b$ with τ_a^b, which he calls the *stress-energy tensor for matter*.

The General Time-Varying Yilmaz Theory

The initial solution derived by Yilmaz applies exactly only when the gravitational field does not vary with time. It is called a *static solution*. The first paper on the Yilmaz theory, which presented this static solution, was published in 1958. After 15 years of intense research, Yilmaz was able to generalize his theory to obtain his *time-varying solution*, which was published in 1973.

A derivation of the time-varying Yilmaz theory is described in the website *Addendum* [3], Chapter 5 and Appendix F. This material is written for the scientist. The time-varying Yilmaz theory is much more complicated than the static solution, yet is still very much easier to apply than the Einstein theory.

One can prove from the general time-varying Yilmaz theory that the simple static solution of the Yilmaz theory gives a very accurate approximation if the gravitational field varies slowly relative to the speed of light. This condition is satisfied in nearly all practical applications. The time varying solution of the Yilmaz theory is needed to describe gravitational waves. (Gravitational waves have also been predicted by the Einstein theory). Except for gravitational waves, the simple static solution of the Yilmaz theory is nearly always more than adequate.

Discussion of the Yilmaz Theory

We have seen that the Yilmaz gravitational theory is a direct extension of the Einstein General theory of Relativity. It applies the principles that Einstein specified in developing his theory.

Yilmaz discovered an exact solution to the relativistic principles that Einstein established. If Einstein had retraced his steps, he probably would have discovered this exact solution that Yilmaz found. This exact solution would have been far more desirable than the gravitational field equation that Einstein chose to specify his theory. Einstein developed his gravitational field equation intuitively, after many frustrating years of searching for an answer.

The derivation of the Einstein gravitational field equation is explained by John A. Peacock in his lengthy scientific book, titled *Cosmological Physics* [6]. This book describes theoretical Big Bang research and includes a detailed discussion of Relativity theory.

Peacock states [6] (p. 19) that the Einstein gravitational field equation *"cannot be derived in any rigorous sense; all that can be done is to follow Einstein and start by thinking about the simplest form such an equation might take."*

On pp. 26-27, Peacock [6] discusses possible *"alternative theories of gravity"*. He asks, *"How certain can we be that Einstein's theory of gravitation is correct?"* He notes that the Einstein theory does not apply in sub-atomic scales, and goes on to say, *"Apart from this restriction, there are no obvious areas of incompleteness. . . Nevertheless, . . it is possible that more accurate experiments will yield discrepancies. Over the years this possibility has motivated many suggestions of alternatives to general relativity."*

Peacock is a strong supporter of the Big Bang theory, which is solidly tied to the Einstein gravitational field equation. Nevertheless, even Peacock admits that the Einstein gravitational field equation was obtained in an intuitive manner, and that a number of alternatives to that equation have been seriously considered by responsible scientists.

In contrast, Yilmaz derived the gravitational field equation of his theory by means of rigorous analyses that applied the principles of the Einstein theory.

The static solution of the Yilmaz theory is remarkably simple even though the theory has a very solid mathematical foundation. As we have seen, the simple static solution is more than adequate for nearly all practical applications.

The Yilmaz theory does not allow the physically impossible black-hole and Big Bang singularities that have been derived from the Einstein gravitational field equation. Einstein strongly opposed the singularity solutions obtained from his theory. He insisted that, ***"singularities do not exist in physical reality"***.

Reason for Opposition to the Yilmaz Theory

With these great advantages of the Yilmaz theory, one might assume that the scientific community would be clamoring for it. This might be true if the Einstein General theory of Relativity were being used in practical applications, but that is rarely the case.

The Einstein theory is the basis for an enormous amount of artificial cosmological research performed over the past 30 years by many hundreds of scientists throughout the world. Computer studies of the Einstein gravitational field equation are the foundation for this research. Because of the extreme mathematical complexity of the Einstein gravitational field equation, sophisticated mathematical techniques have been required to analyze it on the computer.

If the validity of the Yilmaz theory were accepted, it would be clear that the Einstein gravitational field equation is flawed. This finding would make obsolete and irrelevant the enormous cosmological research that has been based on the Einstein equation. The sophisticated computer techniques are not needed to apply the simple Yilmaz theory of gravity.

Therefore, it is not surprising that the army of General Relativity experts that are involved in Big Bang cosmology studies are strongly opposed to the Yilmaz theory.

Consistency with Quantum Mechanics

After presenting his General theory of Relativity, Einstein did little with this theory. He devoted the rest of his life primarily to the task of developing a *unified field theory*. This would have combined into a single theory the concepts of gravitational fields, electromagnetic fields, and atomic nuclear fields. He never succeeded, although he struggled with this task until his last days. An important reason for his failure is that the Einstein gravitational field equation conflicts with quantum mechanics.

Yilmaz has proven that his gravitational field equation is consistent with quantum mechanics. This finding opens great possibilities for relating Relativity theory to quantum field theory, which may eventually lead to Einstein's elusive goal of a *unified field theory*.

Chapter 10

Applying the Einstein and Yilmaz Theories

Relativistic Effects Produced by Gravity

This chapter applies the Einstein and Yilmaz theories to obtain plots of the relativistic effects produced by the gravitational field of a star.

Special Relativity shows that velocity causes a clock to run slower and a measuring rod to contract. In General Relativity, a gravitational field also causes a clock to run slower and a measuring rod to contract. In Special Relativity the speed of light is always constant, but in General Relativity a gravitational field causes the speed of light to decrease. Because of the reduction of clock rate, excited atoms oscillate at lower frequency and this produces a gravitational redshift.

As shown in Table 10-1, these gravitational effects can be calculated from the metric tensor elements. These formulas are obtained from Table E-2 of Appendix E.

Table 10-1: General formulas for relativistic effects due to gravity

clock rate: $\sqrt{[g_{00}]}$ speed of light: $\sqrt{[g_{00}/(-g_{11})]}$
wavelength (λ'/λ): $1/\sqrt{[g_{00}]}$ spatial contraction: $1/\sqrt{[-g_{11}]}$

The following discussion applies the Einstein and Yilmaz theories to specify the gravitational effects of a star. The Einstein theory uses the Schwartzschild solution, which requires that the star have a constant density of matter. With the Yilmaz solution, the density of the star can vary with radius. Consequently the Yilmaz analysis assumes a very much more realistic physical model of a star than does the

Schwartzschild solution. The only restriction of the Yilmaz analysis is that the density of the star must be spherically symmetric.

The Normalized Relativistic Mass m

The gravitational field of a star is proportional to the ratio M/r, where M is the mass of the star and r is the radial distance from the center of the star. To handle this M/r ratio conveniently, a normalized mass (m) is defined as (MG/c^2), as shown in item (1) of Table 10-2. The parameter G is Newton's gravitational constant.

Table 10-2: *Calculating the normalized mass-to-radius ratio at the surface of the sun*

(1) Normalized mass (m)	MG/c^2
(2) Mass of sun (M)	2×10^{30} kilogram
(3) Approximate normalized mass (m) of sun	1.5 kilometer
(4) Radius (r) of sun	700,000 kilometer
(5) Normalized mass-to-radius ratio (m/r) of sun	2.1×10^{-6}

Item (2) of Table 10-2 gives the mass (M) of the sun. Multiplying this mass (M) by (G/c^2) gives the normalized mass (m) of the sun shown in item (3) (This is an approximation; the exact value is 1.475 km.) Item (4) gives the radius (r) of the sun. Dividing item (3) by item (4) gives the (m/r) ratio for our sun shown in item (5).

Since the normalized mass (m) has the units of distance (i.e., km), the (m/r) ratio is a simple number without units. *Normalized mass is a convenient artifice that simplifies our discussion; it does not imply that true mass can be measured in kilometers.*

The (m/r) ratio at the surface of the sun is (1.5 km)/(700,000 km), which is 2.1×10^{-6}, as shown in item (5). Since the factor 10^{-6} means 1/1,000,000, this m/r ratio represents 2.1 parts per million.

The gravitational field at any point outside a star (or our sun) is expressed in terms of the (m/r) ratio at that point, where r is the radial distance measured from the center of the star.

The (m/r) ratio in item (5) of Table 10-2 specifies the gravitational field at the surface of the sun. At any point outside the sun, the radial distance (r) is the distance measured from the center of the sun. The radius of the earth orbit is 150,000,000 km, and so the m/r ratio for the gravitational field of the sun at the location of the earth is equal to (1.5 km)/(150,000,000 km), which is 10^{-8}, or one part in 100 million.

The maximum value of the (m/r) ratio in our solar system occurs at the surface of our sun, where it is only 2.1 parts per million. At the distance of the earth orbit, the m/r ratio is only one part in 100 million. These numbers help to place the following discussion in perspective.

Effect of Gravity on the Speed of Light

Figure 10-1 shows how the speed of light should vary with the gravitational field of a star is accordance with the Einstein and the Yilmaz theories of gravity. The formulas for these plots and the plots in Figs. 10-2 and 10-3 are calculated in Appendix E. They apply the general formulas given in Table 10-1. The plots are shown as solid curves for the Schwartzschild solution of the Einstein theory, and as dashed curves for the Yilmaz solution.

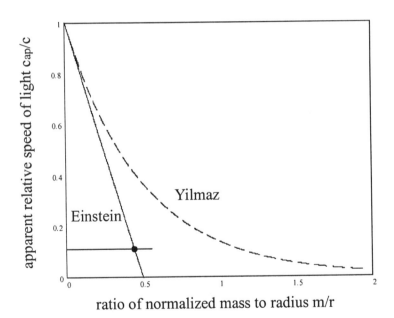

Figure 10-1: Apparent relative speed of light for Schwartzschild Einstein solution and for Yilmaz single-star solution.

In Fig. 10-1, the gravitational field is expressed in terms of the m/r ratio. The maximum value of the m/r ratio in our solar system is only 2.1 parts per million, which is the value at the surface of the sun. Therefore, only the first tiny bit of the graph applies to our solar system. In this initial region, the plots for the Einstein and Yilmaz theories are essentially the same.

Figure 10-1 shows the relative speed of light c_{ap}/c for the two theories, where c_{ap} is the apparent speed of light under the gravitational field, and c is the normal speed of light experienced on earth.

For the Einstein theory, the speed of light goes to zero at m/r equal to ½. For larger values of m/r, the Schwartzschild theory gives an "imaginary" value for pressure inside the star, which means that the Schwartzschild analysis does not have a solution. This point where m/r is ½ is called the *Schwartzschild limit*. Einstein was not concerned about this limit to the Schwartzschild analysis, because it requires a value of m/r that is 240,000 times greater than the maximum value occurring in our solar system. He was interested in applying his theory to practical applications, and so he ignored this limit to the Schwartzschild analysis.

At the point where m/r is equal to ½, the radius r is equal to 2m. This distance (2m) is called the *Schwartzschild radius*. For our sun the Schwartzschild radius is 3 kilometers. If the Schwartzschild radius lies inside the star (as it does for our sun), the Schwartzschild analysis applies. If a star is so compact that the Schwartzschild radius lies outside the star, the Schwartzschild analysis does not yield a solution.

The speed of light plot for the Yilmaz theory in Fig. 10-1 does not go to zero at any value of m/r. Consequently, there is no upper limit to the Yilmaz analysis.

The Black Hole

As was explained earlier, Oppenheimer and Snyder in 1939 showed that they could obtain a solution from the Einstein gravitational field equation when m/r exceeds the Schwartzschild limit of ½, if they assume that the radius of the star decreases with time. The Schwartzschild analysis did not allow a solution above the Schwartzschild limit, because it assumed that the radius of the star remained constant.

Oppenheimer and Snyder concluded from their analysis that if m/r exceeds the Schwartzschild limit, the star must "contract indefinitely". The star should shrink until it becomes a *singularity* having a diameter of zero and an infinite density of matter.

Einstein strongly opposed this *Schwartzschild singularity*. In his rebuttal he insisted that, *"Schwartzschild singularities do not exist in physical reality"*. Einstein realized that a physical singularity would drastically violate our laws of physics. Since Einstein demanded that his theory must be consistent with physical evidence, he knew that a

Schwartzschild singularity would invalidate his General theory of Relativity. Einstein presented an analysis to show that the predicted *Schwartzschild singularity* would not actually occur.

While Einstein was alive, no scientist attempted to dispute Einstein's refutation of the *Schwartzschild singularity*. Nevertheless, the concept became a popular notion in science fiction. If the Schwartzschild limit can be exceeded, the star would be surrounded by a spherical surface over which the speed of light is zero. The surface was called the *event horizon*, because time would theoretically stand still over this surface where the speed of light is zero. The spherical event horizon surface would have the Schwartzschild radius ($r = 2m$).

Light theoretically cannot escape from inside the event horizon surface, and so the star was called a *black hole*. There were many science fiction accounts of the concept of *"falling into a black hole"*. The stories claim that the gravitational force is so great that nothing can oppose the tremendous gravitational pull of a *black hole*.

Although the black hole concept is well known to the general public, it is not generally recognized that the star inside a black hole must "contract indefinitely" until it shrinks to a singularity having zero diameter. Since the star mass does not change as the star shrinks, the density of matter must become infinite.

The dashed plot for the Yilmaz theory in Fig. 10-1 shows that the Yilmaz theory does not allow a black hole. It is clear from the Yilmaz theory that the black hole concept is nothing more than a mathematical defect in the Einstein gravitational field equation. Since the Yilmaz theory is a refinement of the Einstein theory, the Yilmaz theory has proven that the basic principles of the Einstein General theory of Relativity are inconsistent with black holes. ***In agreement with Einstein, Yilmaz has proven that the basic Einstein theory, when properly refined, does not predict physically impossible singularities.***

When computer studies of the Einstein gravitational field equation were implemented in the 1960's, about a decade after Einstein's death, it was concluded that the Einstein theory does indeed predict a singularity. Consequently the black hole concept was officially endorsed by the scientific community. Since the Einstein theory predicts the existence of a black hole, it was concluded that a black hole must actually exist "in physical reality".

Let us reexamine the following quotation given earlier in Chapter 4, which was obtained from Filkin's book *Stephen Hawking's Universe* [19] (p. 104):

156 *The Scientific Story of Creation*

"Stephen (Hawking) and Roger Penrose published a paper in 1970 which proved that, if Einstein's mathematics were correct, a singularity had to result from a black hole, and had to exist at the start of the universe. - - - The paper argued that if relativity as explained by Einstein is correct — and all of the evidence from observation seems to keep confirming it — then the universe must have started with a big bang explosion out of a singularity. The equations do not allow an alternative."

Hawking and Penrose are insisting that singularities must physically exist simply because they are predicted by the Einstein theory. They claim that we should believe in the reality of singularities because the Einstein theory has been verified by observational evidence. However, this observational evidence has been obtained from experiments performed within our solar system, where the gravitational field is so weak that m/r is no greater than 2 parts per million. Figure 10-1 shows that the plots for the Einstein and the Yilmaz theory are essentially the same at these extremely low values of m/r.

Hawking and Penrose maintain that experimental verification of the Einstein theory proves that the Einstein theory must hold in the region above the Schwartzschild limit, where the m/r ratio is more than one-quarter million times greater than has occurred in these experiments. *This is an extremely poor argument for justifying the claim that physically impossible singularities actually occur in nature.*

Second Limit to the Schwartzschild Solution

Although the Schwartzschild analysis yields a real solution up to the Schwartzschild limit where m/r is equal to ½, a physically impossible condition actually occurs at a lower value of m/r. If m/r is greater than 4/9 but less than ½, the Schwartzschild analysis predicts that the pressure inside the star should be infinite over a spherical surface. This surface progresses from the center of the star to the circumference as the m/r ratio is increased from 4/9 to ½.

Since infinite pressure is not physically possible, it has been concluded that a star would become unstable if m/r exceeds 4/9, and would collapse to form a black hole. The dot on the plot for the Einstein theory in Fig. 10-1 shows this upper limit to the m/r ratio. Presumably a star cannot exist in a stable condition if m/r exceeds this value. It is generally assumed that if m/r exceeds 4/9 the star must collapse to become a black hole.

Gravitational Effect on Distance and Clock Rate

A gravitational field causes a distance to contract and a clock to run slower. Figure 10-2 shows the contraction of dimension and the reduction of clock rate that are predicted by the Einstein and the Yilmaz theories. For these examples, the relative clock rate and the spatial contraction ratio are equal.

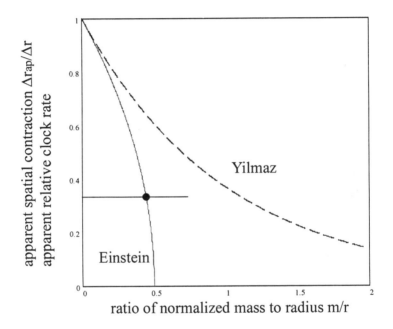

Figure 10-2: Apparent spatial contraction and relative clock rate for Einstein Schwartzschild solution and Yilmaz theory

We have seen that the Einstein theory cannot allow a value for m/r in excess of 4/9. Figure 10-2 shows that the minimum possible value for the spatial contraction ratio or relative clock rate is 1/3, which occurs at m/r equal to 4/9. This minimum value for the Einstein theory is indicated by a dot in the figure.

In contrast, the spatial contraction ratio and relative clock rate for the Yilmaz theory (the dashed plot) does not have a lower limit.

Effect of Gravitational Field on Wavelength

The decrease of clock rate due to a gravitational field given in Fig 10-2 shows that gravity causes the period of a clock to increase. The wavelength of a light wave is equivalent to a clock period. Consequently a gravitational field causes a wavelength to increase by an amount inversely proportional to the decrease of clock rate. Figure 10-3 shows how the wavelength increases with increasing m/r ratio.

Figure 10-3 shows the wavelength ratio λ'/λ where λ is the normal wavelength and λ' is the observed wavelength. Since the maximum allowable value of m/r for the Einstein theory is 4/9, Fig. 10-3 shows that the maximum value of the λ'/λ wavelength ratio for the Einstein theory is 3. This maximum value is shown as a dot in the figure.

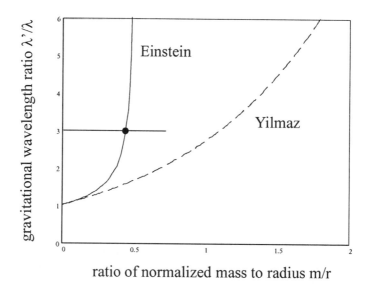

Figure 10-3: Wavelength ratio due to gravity for Einstein Schwartzschild solution and for Yilmaz theory

The observed wavelength λ' is equal to $(\lambda + \Delta\lambda)$, where $\Delta\lambda$ is the increase in wavelength above the normal wavelength λ. Redshift is defined as the ratio $\Delta\lambda/\lambda$ and so is equal to $(\lambda'/\lambda - 1)$. Figure 10-3 shows that the maximum λ'/λ wavelength ratio due to gravity that can be predicted by the Einstein theory is 3. Consequently the maximum $\Delta\lambda/\lambda$ gravitational redshift that the Einstein theory can predict is 2.

10. Applying the Einstein and Yilmaz Theories 159

In Fig. 10-3, the dashed curve of λ'/λ for the Yilmaz theory does not have an upper limit. Therefore the Yilmaz theory shows that there is no upper limit to the value of gravitational redshift.

The Quasar Redshift

When the extreme redshift of the quasar was discovered in 1963, it was recognized by astronomers that the quasar would become a much more reasonable star if the extreme wavelength shift could be attributed to an intense gravitational field rather than to an extreme velocity. We have seen that the Einstein theory cannot predict a gravitational redshift greater than 2. However, this did not initially rule out gravitational redshift, because the early quasars had redshifts less than 2.

Chandrasekhar [34] proved from an analysis of the Einstein theory that a star should exhibit strong radial oscillations unless the m/r ratio at its surface is much less than 4/9. It was assumed that these oscillations would cause a star to collapse into a black hole. Consequently, it was believed that the maximum gravitational redshift that could be achieved by a stable star is much less than 2.

A second argument against gravitational redshift is that quasar spectra display "forbidden" spectral lines of oxygen and neon. Forbidden spectral lines are never encountered on earth and, except for quasars, are only observed in a gaseous nebula. A gaseous nebula is a very thin cloud of gas in space that is heated to a temperature of about 10,000 degrees Kelvin by the radiation from nearby stars. It was concluded that a star that has sufficient density to exhibit a large gravitational redshift would be too compact to emit forbidden spectral lines.

With these two arguments, the gravitational redshift explanation for the quasar spectrum was soundly rejected in the early quasar studies. Nevertheless, a fresh examination of the evidence shows that gravitational redshift remains a strong possibility for explaining the quasar redshift.

The argument raised by Chandrasekhar [34] can be easily refuted. The oscillation condition derived by Chandrasekhar applies only to the Einstein theory. The Yilmaz theory does not exhibit any such problem. Consequently the Yilmaz theory can readily predict all of the quasar redshifts as gravitational effects. The maximum quasar redshift found to date is about 5, which corresponds to a λ'/λ wavelength ratio of 6. [37]

The issue of forbidden spectral lines is discussed in Chapter 11, which explores the quasar enigma.

Chapter 11

The Quasar

Summary of Quasar Characteristics

As explained in Chapter 4, the quasar is a perplexing enigma. A quasar is a star-like object that looks like a point of light to the telescope and has an extremely large redshift in its spectrum. If this redshift is a Doppler velocity effect, a quasar must be at an enormous distance, billions of light years away. At such a distance, it must radiate an unbelievable amount of power for astronomers to receive the light that is measured. Yet the power radiated by many galaxies varies rapidly, which indicates that these quasars are small.

For example, Appendix B examines the data for quasar 3C48, which was one of the first two quasars to be studied. If this quasar is at the distance indicated by its redshift (which is 3.6 billion light years) it would have to radiate a power equivalent to 1000 billion suns, which is 100 times the total power radiated from our Milky Way galaxy.

The power from 3C48 varied by 40 percent over a period of 600 days. This indicates that at least 40 percent of the power (the radiation from 400 billion suns) must come from a body that is no more than 600 light-days (or 1.6 light years) thick. This enormous power is being radiated from an extremely small volume.

Soon quasars were discovered with much larger redshifts and with much more rapid variation of power. Redshifts up to 5 have been discovered, which means that the observed wavelength is 6 times the normal wavelength. Quasars were found that vary by 2-to-1 in power over periods of months, weeks, days, and even hours. This indicates that some quasars are no larger than our solar system. [37]

Quasar Observations of Halton Arp

Because of the unbelievable properties attributed to quasars, the noted astronomer, Halton Arp, suspected that they may be much closer than was being assumed. He began to make observations of quasars to obtain direct astronomical estimates of their distances.

Arp found many quasars with images that are very close to galaxies having much smaller redshifts. If the image of one quasar is very close to that of a galaxy, this might be a chance relationship. The quasar might be billions of light years beyond the galaxy. However, if two or more quasars appear to be close to a galaxy, the probability of a chance relationship is remote. We ask the question, "What is the probability that two or more quasar images would fall this close to an arbitrary direction in space?" Probability considerations indicate that it is highly unlikely that the quasars are not physically close to the associated galaxy.

In Halton Arp's study of quasars, he also found many cases of filaments that directly connect quasars to galaxies having much smaller redshifts. These observations suggest that the quasar was ejected from the associated galaxy by a supernova explosion. There are several cases of a filament connecting a quasar to a galaxy, and another opposing filament on the other side of the galaxy. This suggests that the opposing filament is the reaction from a supernova explosion that ejected the quasar from the galaxy.

The quasar observations by Arp were opposed by the general astronomical community, because they did not agree with the accepted dogma that quasars are billions of light years away. He found it very difficult to get his findings published. Some journals rejected his work, and papers were often held up for years by referees. Finally in 1984, the committee that controls observation time at the Palomar and Mount Wilson Observatories refused to allow Arp to use these facilities.

Dr. Halton Arp had performed distinguished research at Palomar and Mount Wilson Observatories since he received his PhD degree in 1953. He was president of the Astronomical Society of the Pacific from 1980 to 1983, and received awards from the American Astronomical Society, the American Association for the Advancement of Science, and the Alexander von Humbolt Senior Scientist Award. After being denied research facilities in California, he was forced to move to Germany to continue his career, where he joined the Max Planck Institute for Physics and Astrophysics in Munich.

Arp presented his quasar observations up to the time of his expulsion in his 1987 book, *Quasars, Redshifts, and Controversies*. [16] In his later 1998 book, *Seeing Red* [17], he also includes the extensive observations on quasars that he has made since he was forced to move to Germany.

Arp's 1998 book gives overwhelming evidence that quasars are very much closer than is generally assumed. Nevertheless, the leaders in astronomy completely ignore this evidence and continue to insist that the conventional quasar dogma is correct.

Statistical Evidence Given by Halton Arp

The books [16, 17] by Halton Arp give highly convincing photographs showing many examples of filaments that connect quasars to galaxies having much lower redshifts. These photographs prove that the velocities of many quasars are much less than is indicated by their redshift values. This means that quasars are much closer than is generally assumed, and are radiating very much less power.

Let us now examine the strong statistical evidence that Arp has presented, proving that some quasars are physically related to galaxies having much smaller redshifts. Consider Table 11-1. The data in columns (1) to (5) are obtained from Table 1-1 of Arp [16] (p. 13). Table 11-1 gives data for quasars that are very close to four galaxies. Two of these galaxies have two nearby quasars, and two have three nearby quasars.

Column (1) in Table 11-1 lists the galaxies (in **bold** letters) and the associated quasars. Column (2) gives the redshifts for the galaxies and the quasars, which show that the quasar redshifts are very much larger than those of the galaxies.

Column (3) gives the angular separation of each quasar from the center of the associated galaxy. For the galaxies, the values in parentheses are the approximate angular diameters of the galaxies, which the author read from Arp's photographs. The values in parentheses allow the reader to compare a quasar distance to the size of the associated galaxy.

Based on each angular separation given in column (3), Arp calculated the corresponding probability that a quasar would be this close to an arbitrary direction in space, assuming that quasars are spaced uniformly. As will be explained, this probability calculation considers only the known quasars that have equal or greater brightness than the particular quasar. The probability values are shown in column (4).

Table 11-1: Galaxies with multiple quasars found by Arp, and power radiated, if quasar is at distance of associated galaxy

galaxy quasar	redshift $\Delta\lambda/\lambda$	angular separation	probability	magnitude	power in suns
NGC622	0.018	(80 sec)*			
UB1	0.91	71 sec	0.001	18.5	150 million
BS01	1.46	73 sec	0.02	20.2	31 million
NGC470	0.009	(134 sec)*			
68	1.88	95 sec	0.015	19.9	11 million
68D	1.53	95 sec	0.002	18.2	50 million
NGC1073	0.004	(172 sec)*			
BS01	1.94	104 sec	0.01	19.8	2.3 million
BS02	0.60	117 sec	0.006	18.9	5.2 million
BS03	1.40	84 sec	0.02	20.0	1.9 million
NGC3842	0.020	(64 sec)*			
QS01	0.34	73 sec	0.003	19.0	120 million
QS02	0.95	59 sec	0.002	19.0	120 million
QS03	2.20	73 sec	0.01	21.0	19 million
(1)	(2)	(3)	(4)	(5)	(6)

* Approximate angular diameter of galaxy image

Table 11-2: Parameters of galaxies listed in Table 11-1

galaxy	redshift	distance MLyr	diameter angular	diameter light years
NGC622	0.018	216	80 sec	83,000
NGC470	0.009	108	134 sec	70,000
NGC1073	0.004	48	172 sec	40,000
NGC3842	0.020	240	64 sec	74,000
(1)	(2)	(3)	(4)	(5)

164 *The Scientific Story of Creation*

For galaxy NGC 3842, the three quasars have the following probabilities of being accidental relationships: 0.003, 0.002, and 0.01. If we multiply these three probabilities, we have a total probability of 60×10^{-9}, or 60 chances in one billion, that all three quasars are unrelated to the associated galaxy. Arp states conservatively that the chances for this are one in a million.

But this is just for one galaxy. The probability that all of the galaxy-quasar relationships shown in Table 11-1 are accidental is essentially zero. Therefore Table 11-1 gives very strong evidence that at least some quasars are physically close to galaxies of much lower redshifts.

Besides this very strong evidence, Arp has presented other statistical data indicating that the redshift of a quasar does not specify its velocity.

Principle for Calculating Probabilities

Since the implications of the probability values given in Table 11-1 are very important, let us see how they were calculated. The probability values in column (4) are based on the angular separation values in column (3) and on the magnitude values in column (5).

First, consider a circle passing through a quasar image that has its center at the center of the galaxy image. The area of this circle is expressed in square degrees. This area is multiplied by the average number of known quasars per square degree that are of equal or greater brightness than the particular quasar. The resultant product is the probability value in column (4).

The area of the circle in square degrees is equal to

Area in square degrees $= \pi \, (\phi_{sec}/3600)^2$

The parameter ϕ_{sec} is the angular separation value in seconds that is given in column (3) for the particular quasar. The probability calculation is based on the average number of known quasars that have equal or lower magnitude than the particular quasar.

For example, assume that the angular separation is 65 seconds. The above formula shows that the circular area is 0.0010 square degrees. Assume that the quasar has a magnitude of 20. Arp has reported that catalogues of known quasars give quasar densities of 6 to 10 quasars per square degree for quasars of magnitude 20 and below. (Remember that the lower the magnitude, the brighter is the star.) Arp conservatively uses the higher density of 10 quasars per square degree. This is multiplied by the circular area of 0.0010 square degrees to yield a probability of 0.010 for that quasar.

How Much Power Does a Quasar Radiate?

The data of Table 11-1 can be used to calculate the power radiated by an individual quasar by assuming that it is at the same distances as its associated galaxy. The distances of the four galaxies are calculated in Table 11-2.

Column (2) in Table 11-2 gives the redshift values for the galaxies shown in column (2) of Table 11-1. These redshift values are multiplied by 12 billion light years (12,000 MLyr) to obtain the corresponding quasar distances in column (3). This calculation assumes a Hubble constant or 25 km/sec per million light years, which corresponds to an apparent universe age of 12 billion years. (Elsewhere, this book has usually assumed a Hubble constant of 20 km/sec per million light years, which corresponds to an apparent universe age of 15 billion years.)

Based on the galaxy distances in column (3) of Table 11-2 and the magnitudes of the associated quasars in column (5) of Table 11-1, one can calculate the power emitted by each quasar, which is shown in column (6). The power is expressed in terms of equivalent suns. The details of this calculation are given in *Believe* [1], Appendix F.

Column (6) in Table 11-1 shows that the power radiated from the quasars varies from 2 million suns to 150 million suns. *This indicates that quasars are radiating huge amounts of power even though the power levels are tremendously smaller than is normally assumed.*

Diameters of the Associated Galaxies

Table 11-1 gives the angular diameters of the four galaxies, which are shown in parentheses in column (3). These values were read by the author fom Arp's photographs. These values are also shown in column (4) of Table 11-2. By combining these angular values with the galaxy distances in column (3), the corresponding galaxy diameters in light years are calculated, as shown in column (5).

Column (5) of Table 11-2 shows that the galaxy diameters range from 40,000 to 83,000 light years. For comparison, the diameter of our Milky Way galaxy is 100,000 light years. Since the Milky Way is a large galaxy, these diameters seem to be reasonable. This indicates that our calculated galaxy distances should be roughly correct.

Galaxies with Intrinsic Redshift

Halton Arp's astronomical observations have provided very strong evidence that, for at least some quasars, the redshift of a quasar does not specify its velocity or its distance. In Arp's terminology, a quasar has an "intrinsic redshift" component that is unrelated to its velocity.

Arp has also found strong evidence that galaxies can have intrinsic redshift. He has recorded many examples of a galaxy that is directly connected by a filament structure to a larger galaxy having much lower redshift. The evidence often suggests that the smaller galaxy was ejected from the larger galaxy. In *Seeing Red* [17], Arp describes several examples of galaxies with intrinsic redshift

Possible Explanations for Intrinsic Redshift

Arp has provided strong evidence of quasars and galaxies having intrinsic redshifts that are unrelated to their velocities and so are not Doppler effects. What is causing intrinsic redshift? We have found the following two redshift effects that are strong candidates for explaining intrinsic quasar redshift:

(1) Gravitational redshift, discussed in Chapter 10.
(2) The redshift effect of Paul Marmet, discussed in Appendix A.

The Marmet effect (2) can also explain intrinsic galaxy redshift.

As shown in Chapter 10, the amount of gravitational redshift that can be predicted by the Einstein theory is very limited. However, the Yilmaz theory can predict a very large gravitational redshift.

The Marmet redshift effect, which is described in Appendix A, predicts that a cloud of hydrogen gas can produce a redshift. The redshift is proportional to the density of the gas and to the path length. Many gaseous nebulae have gas densities of 100,000 molecules per cubic centimeter. A gas cloud with this density should produce an intrinsic redshift of 0.2 for every 1000 light years of cloud thickness.

Intrinsic Redshift of Quasar 3C48

Appendix B examines quasar 3C48, which was one of the first two quasars studied. The image of this quasar is surrounded by a faint nebulous object with an angular length of 12 arc seconds. Redshift measurements of this nebula have found its redshift to be constant across the nebula and to be within 0.001 of the quasar redshift.

This finding is inconsistent with the gravitational redshift hypothesis, because a body would have to be very compact to have a high gravitational redshift. On the other hand, Appendix B shows that the Marmet redshift effect seems to give an acceptable explanation of the intrinsic redshift of Quasar 3C48 and the redshift of the associated nebulous object.

The Implications of Forbidden Spectral Lines

In the analysis of data for quasar 3C48, an important argument against gravitational redshift as the cause of the quasar redshift was that the quasar spectrum displays "forbidden" spectral lines of oxygen and neon. Forbidden spectral lines are never encountered on earth and, except for quasars, are only observed in the light from gaseous nebulae. A gaseous nebula is a very thin cloud of gas in space that is heated to a temperature of about 10,000 degrees Kelvin by the radiation from nearby stars. It was concluded that a star that has sufficient density to exhibit a large gravitational redshift would be too compact to emit forbidden spectral lines.

In their analysis of 3C48, Greenstein and Schmidt concluded that the quasar was at a distance of 3.6 billion light years, and that the forbidden lines are being radiated from a gas cloud 65 light years in diameter. The density of the gas cloud was estimated to have ionized gas with a density of 30,000 electrons per cubic centimeter.

The power level of the 3C48 radiation varied by 40 percent over a period of 1.6 years, which indicates that the primary body of 3C48 cannot be more than a few light years thick. However, no measurements were made of the time variation of power in the forbidden spectral lines. If these forbidden lines vary in power along with the general power level of 3C48, then the explanation for these forbidden spectral lines would be unacceptable. The thickness of the gas cloud would have to be very much smaller than 65 light years.

It is clear that more measurements of forbidden spectral lines of quasars are needed, particularly their time variation.

It has been assumed that the forbidden spectral lines in quasar spectra must come from large nebulous clouds of diffuse gas, because that is the only other source of forbidden spectral lines that has been observed. However, until astronomers can achieve explanations of the quasar spectra that are consistent with physical evidence, other possible explanations for forbidden spectral lines should be considered.

If a quasar is a highly compact body, it would have physical conditions in its atmosphere that are radically different from those encountered on earth. It would have an extremely high gravitational gradient and would probably have a very large magnetic field. Can forbidden spectral lines be generated under these severe physical conditions?

Explanations for Intrinsic Quasar Redshifts

Although gravitational redshift does not appear to be an acceptable explanation for the redshift of quasar 3C48, it may well be a good explanation for other quasars. Both the Marmet redshift effect and gravitational redshift should be considered as serious candidates for explaining the quasar spectra.

Cosmologists have proposed physically impossible concepts to explain the quasar spectra. A common postulate is that a quasar is the interaction between two physically impossible black holes. When scientific objectivity returns to astronomy, and all of the quasar data are carefully considered, it seems likely that explanations of quasars will emerge that are physically reasonable.

Confusion in Quasar Research

Authorities in the field of astronomy have rejected the excellent astronomical data obtained by Halton Arp, because it is inconsistent with accepted dogma. In order to suppress his results, they have denied this highly respected astronomer the use of the Palomar and Mount Wilson Observatories. Since astronomical research today is very selective of the data that it considers, it is incapable of deriving conclusions that have scientific validity. *In this scientific chaos, we have returned to the philosophy of medieval astronomy, where truth is established by the opinions of an intellectual elite.*

Chapter 12

Evidence against the Big Bang

For many years the official position of the field of astronomy has been to treat the Big Bang theory as fact, even though, as this book has shown, there are strong reasons to doubt the theory. On the other hand, many voices have been raised against the Big Bang. Let us consider some of them.

The Editorial of Geoffrey Burbidge

The case against the Big Bang was expressed eloquently by Professor Geoffrey Burbidge in an editorial article of the February 1992 *Scientific American* [27]. Professor Burbidge is the former director of the Kitt Peak National Observatory, and is presently Professor of Astrophysics at the University of California in San Diego. Prof. Burbidge began his editorial with:

> *"Big bang cosmology is probably as widely believed as has been any theory of the universe in the history of Western civilization. It rests, however, on many untested, and in some cases untestable, assumptions. Indeed, big bang cosmology has become a bandwagon of thought that reflects faith as much as objective truth."*

He went on to say

> *"Younger cosmologists are even more intolerant of departures from the big bang faith than their more senior colleagues are. Worst of all, astronomical textbooks no longer treat cosmology as an open subject. Instead the authors take the attitude that the correct theory has been found."*

Burbidge then explained the basic reasons for the bandwagon mentality:

> *"Powerful mechanisms encourage this conformity. Scientific advances depend on the availability of funding, equipment, and journals in which to publish. Access to these resources is granted through a peer review process. Those of us who have been around long know that peer review and the refereeing of papers have become a form of censorship. It is extraordinarily difficult to get financial support or viewing time on a telescope unless one writes a proposal that follows the party line."*

> *"A few years back, Halton C. Arp was denied telescope time at Mount Wilson and Palomar Observatories because his observing program had found, and continued to find, evidence contrary to standard cosmology."*

> *"Unorthodox papers often are denied publication for years or are blocked by referees. The same attitude applies to academic positions. I would wager that no young researcher would be willing to jeopardize his or her scientific career by writing an essay such as this."*

Burbidge concluded with the following:

> *"The big bang ultimately reflects some cosmologist's search for a creation and a beginning. This search properly lies in the realm of metaphysics, not science."*

Eric Lerner and Nobel Laureate Hannes Alfven

In 1991, Eric Lerner wrote the book, *The Big Bang Never Happened* [18]. He was strongly supported in this book by Nobel laureate Hannes Alfven (1908-1995), who was the father of modern plasma physics. The stimulus for the book was the strong opposition that Hannes Alfven and other plasma physicists had received in their attempts to publish papers in astronomical journals that related plasma physics to cosmology.

There is abundant evidence that plasma physics effects have greatly influenced the development of our solar system, and have caused the rotation of galaxies and stars. Nevertheless, these papers have been continually rejected by the Big-Bang establishment, which controls

12. Evidence against the Big Bang 171

astronomical literature and research funds.

Lerner [18] gives a detailed criticism of Big Bang research, which is only briefly summarized here. His book is highly recommended to show the serious problems in the field of astronomy today.

The Big Bang Age Dilemma

One of the most serious weaknesses of the Big Bang theory is the universe age dilemma. According to the Big Bang theory, the age of the whole universe can be no greater than about 15 billion years, yet there is evidence that some stars in our galaxy are nearly 15 billion years old. An article in the May 2001 *Scientific American* described recent studies of this issue, and stated (page 53) that the age of the oldest stars is 13 billion years, with an uncertainty of ±1.5 billion years.

It has often appeared that some stars are older than our universe. Although recent Big Bang studies have disputed that claim, there is little margin in the calculations. The development of stars and galaxies must have proceeded in an extremely efficient manner in order for our whole universe to have been formed within the 15 billion year maximum age of our universe.

The Big-Bang age dilemma is actually much worse than this discussion suggests. As explained by Eric Lerner [18] (pps 15-32), recent astronomical studies have shown that the universe is not at all uniform. In 1986, Brent Tulley, a University of Hawaii astronomer, (assisted by J. R. Fischer) found that almost all galaxies within 1.5 billion light years are concentrated into huge ribbons, typically one billion light years long, 300 million light years wide, and 100 million light years thick. These ribbons contain curling filaments, a few million light years thick, extending for hundreds of millions of light years.

This study evolved from a mapping of individual galaxies out to 100 million light years. In making this map, it was assumed that the redshift of a galaxy specifies its distance. From this map Tulley and Fischer found (with about 20 exceptions) that all of the thousands of galaxies are concentrated into filaments, a few million light years across. These filaments extend for hundreds of millions of light years, beyond the limits of the map. While performing a later mapping study, astronomer Margaret Haynes concluded after examining the curling galactic filaments, *"The universe is just a bowl of spaghetti."*

To expand his study to 1.5 billion light years, Tulley mapped clusters of galaxies, because there are millions of individual galaxies within that range, too numerous to be mapped individually. From this cluster map Tulley discovered his huge ribbons.

About 1990, Tulley's findings were confirmed by several astronomer teams. The most dramatic is by Margaret J. Geller and John P. Huchra of the Harvard Smithsonian Center for Astrophysics, who are mapping individual galaxies out to 600 million light years, about 200 times as many as in the Tulley-Fischer map of individual galaxies. In their preliminary results, they displayed what they call the "Great Wall", a huge ribbon of galaxies stretching across the region mapped, a distance of 700 million light years. This ribbon, which is 200 million light years wide and 20 million light years thick, corresponds closely to a ribbon mapped by Tulley using clusters of galaxies. Galaxy density inside the ribbon is 25 times greater than outside.

These results seriously contradict the Big Bang theory in two ways: (1) it would probably take at least 150 billion years to form these gigantic structures; and (2) the Big Bang theory predicts a very uniform universe, not the spaghetti-like and ribbon-like structures that are observed.

Lerner [18] (page 23) explains the universe age problem. Except for the general Hubble expansion of the universe, the maximum local velocity of all galaxies is less than 1000 km/sec, which is 1/300 of the speed of light. Since the time of the postulated Big Bang, a galaxy could move only a distance of 15 billion light years divided by 300, which is 50 million light years. However, if the universe were uniform after the Big Bang, galaxies would have had to move 270 million light years to form the huge ribbons. The age discrepancy is worse than these numbers suggest, because time must be allowed for a galaxy to accelerate and decelerate.

Mythological Philosophy of Big Bang Research

The cosmic microwave radiation, which is used as primary evidence to support the Big Bang theory, is actually a serious liability to the theory, because it is far too smooth and uniform. It follows the theoretical blackbody spectrum to very high accuracy, and the radiation arrives with high uniformity from all directions. If this microwave radiation is the cooled relic of light radiated from the very hot matter existing 300,000 years after the Big Bang, the universe must have been extremely uniform at the time the energy was radiated. How did the universe change from that extremely uniform state to the spaghetti-like and ribbon-like universe of today, during the short time following the Big Bang? The 15 billion year age of the universe claimed by Big Bang

12. Evidence against the Big Bang   173

proponents is at least 10 times too short.

Let us ask the more realistic question: What process could cause galaxies to arrange themselves into these spaghetti-like and ribbon-like structures? To this question, Lerner [18] (pps. 39-49) gives a clear scientific answer. For many years, Hannes Alfven, a Swedish Nobel laureate and virtual founder of modern plasma physics, has proposed that electrical currents flowing through the ionized plasma of space produce strong magnetic forces that greatly affect the development of galaxies.

The thin gas of space is ionized, producing what is called plasma. This means that electrons are separated from the nuclei of the atoms, and so can move freely through the plasma to form electric currents. Although the electric current flowing through a square meter of area is very small, an enormous current can flow through the huge area associated with a typical star, which is several light years wide. The electric current flowing through the ionized plasma of space can generate a magnetic field of very high energy, which interacts with the magnetic field of the star to alter the motion of the star. The electric and magnetic fields of the plasma currents tend to produce vortices that could cause the rotation and spiral shape of a galaxy. Over intergalactic distances, the electric and magnetic fields could create the spaghetti-like filaments into which galaxies are arranged.

Alfven has shown in laboratory experiments that instabilities cause plasma currents to form themselves into swirling electrical currents that twist relative to one another like the strands of a rope. The effects of such currents are seen in the aurora, or "northern lights". They are also seen in a gaseous nebula, which is a mass of heated plasma surrounding a group of stars. On an intergalactic scale, plasma electric currents could produce the filament arrangement of galaxies, and the larger ribbon-like structures.

Much more information is given by Lerner [18] concerning the tremendous cosmological implications of plasma physics, which has been pioneered by Nobel laureate Hannes Alfven.

The Big Bang cosmologists have ignored or dismissed plasma theory, and few have even bothered to read about it. The well-known cosmologist, P. James E. Peebles (called the "father of modern cosmology" by *Scientific American*) stated that Alfven's ideas are *"just silly"*. His colleague at Princeton, Jeremiah Ostriker, commented, *"There is no observational evidence that I know of that indicates electric and magnetic forces are important on cosmological scales."*

Alfven, as well as lesser-known plasma physicists, have repeatedly had their papers rejected by astrophysical journals because they contradict Big Bang wisdom. Alfven commented, *"I think the Catholic Church was blamed too much for the case of Galileo — he was just a victim of peer review".*

As evidence against the Big Bang has mounted, Big Bang cosmologists have shrugged it off, and have proceeded to devise more and more elaborate ad-hoc theories to bypass the evidence. Joseph Silk, who has written three books on the Big Bang, stated flatly

> *"It is impossible that the big bang is wrong. Perhaps we'll have to make it more complicated to cover the observations, but it is hard to think of what observations could refute the theory itself."*

Lerner [18] (page 54) responded to this with the following, which must also have reflected the convictions of Nobel laureate Alfven, who helped Lerner greatly in his book:

> *"This attitude is not at all typical of the rest of science, or even of the rest of physics. In other branches of physics, the multiplication of unsupported entities to cover up a theory's failure would not be tolerated. The ability of a scientific theory to be refuted is the key criterion that distinguishes a science from metaphysics. If a theory cannot be refuted, if there are no observations that could disprove it, then nothing can prove it — it cannot predict anything; it is a worthless myth."*

Lerner (p. 56) quotes the following words by Alfven on the myth issue:

> *"The cosmology of today is based on the same mythological views as that of the medieval astronomers, not on the scientific traditions of Kepler and Galileo."*

Big Bang theorists treat the Einstein gravitational field equation as their ultimate truth (their infallible "Bible"). As reported by Lerner (p. 163), cosmology theorist George Field stated his Big Bang philosophy as:

> *"I believe the best method is to start with exact theories, like Einstein's, and derive results from them."*

This philosophy of modern cosmology sharply contrasts with legitimate science, which demands consistency between theory and observation. It is the philosophy of mythology, not of science. Lerner (p. 162) states:

"Entire careers in cosmology have now been built on theories that have never been subject to observational tests, or have failed such tests and have been retained nonetheless."

As Lerner (p. 127) points out, the mythology of modern cosmology is based on the "myth of Einstein". He explains Alfven's concepts with:

"It is quite ironic that the triumph of science [from relativity theory] led to the resurgence of myth. The most unfortunate effect of the Einstein myth is the enshrinement of the belief, rejected for four hundred years, that science is incomprehensible, that only an initiated priesthood can fathom its mysteries."

Lerner quotes the following words of Alfven:

"The people were told that the true nature of the physical world could not be understood except by Einstein and a few other geniuses who were able to think in four dimensions. Science was something to believe in, not something that should be understood. Soon the best sellers among the popular science books became those that presented scientific results as insults to common sense. One of the consequences was that the boundary between science and pseudoscience began to be erased. To most people, it was increasingly difficult to find any difference between science and science fiction."

Since the start of the wide acceptance of the Big Bang theory about 1970, cosmology theorists have been struggling with its many serious conflicts with observational evidence. Lerner [18] (pps 150-163) gives an excellent discussion of the increasingly complicated, bizarre, and arbitrary hypotheses that have been developed since 1965 to accommodate the more and more evident conflicts of the Big Bang theory with observed data.

Effect of the Computer on Cosmological Studies

Albert Einstein died in 1955, which was before computers became generally available. Unless the physical model is very simple, the equations of General Relativity can result in millions of terms, and so cannot be solved analytically. Consequently, Einstein had to limit his General Relativity studies to very simple cases.

When powerful computers became widely available in the mid 1960's, many physicists, mathematicians, and engineers in academic positions began to perform computer studies of General Relativity. With computers, the formidable equations of General Relativity could now be solved in a manner unheard of in Einstein's day.

There was strong economic incentive to perform these studies. Universities are all searching for prestige, which is gauged primarily by publications. The motto among university professors is "Publish or Perish". Since the theory of Albert Einstein is treated with awe, anyone who can obtain a publication based on the Einstein General theory of Relativity is sure to achieve strong commendation from his superiors.

As we have seen, the basic theory of Relativity that Einstein presented in 1905, called *Special Relativity*, has wide applicability. However, the much more complicated *General Theory of Relativity* that Einstein published in 1916 has very limited practical application. Einstein developed his General theory to provide a theoretical foundation for Special Relativity.

Special Relativity does not apply exactly under conditions of acceleration or in a gravitational field, but it yields very accurate approximations in nearly all experiments encountered within our solar system. In those rare cases within our solar system where General Relativity is needed, a simple physical model can be used, because the gravitational field within our solar system is weak.

For these reasons, there is very little need for computer studies of General Relativity that are related to experiments performed within our solar system. The only meaningful area where these computer studies can be applied is cosmology. Therefore, starting in the mid 1960's there was an enormous increase in theoretical Big Bang cosmology research, which were based on computer analyses of the Einstein General theory of Relativity.

Eric Lerner [18] (pps. 253-154) described the tremendous increase of theoretical Big Bang research as follows:

> As Lerner explains, economic forces caused an enormous expansion of theoretical cosmology research. The annual number of cosmology papers skyrocketed from 60 in 1965 to 500 in 1980, and the increase was almost entirely in purely theoretical work. In 1980, 95 percent of these papers were devoted to various mathematical models, such as the *"Bianchi type XI universe"*.

In Lerner's words: during the 1970's, *"the field of cosmology was transformed from a small group of squabbling theorists trying to develop theories that would match observations, to a huge phalanx of hundreds of researchers, virtually all united in their basic assumptions, and preoccupied with the mathematical nuances of the underlying theory."* This enormous theoretical effort in cosmology has continued to this day, and international conferences on cosmology are now held about once a month to discuss the latest hot Big Bang model.

Quasar Studies by Astronomer Halton Arp

We have discussed the case of Halton Arp. He had performed distinguished research at Palomar and Mount Wilson Observatories since he received his PhD degree in 1953. He was president of the Astronomical Society of the Pacific from 1980 to 1983, and received awards from the American Astronomical Society, the American Association for the Advancement of Science, and the Alexander von Humbolt Senior Scientist Award.

One of Halton Arp's best-known publications is his *Atlas of Peculiar Galaxies*. This is a detailed study of the many different types of galaxies, showing that some galaxies have characteristics that are radically different from our own Milky Way galaxy.

Starting in the mid 1960's, Arp turned his attention to the newly discovered quasars. As explained in Chapter 11, Halton Arp found strong statistical evidence of quasars that are parts of nearby galaxies. Besides this strong statistical evidence, he also found many examples of filament structures directly connecting quasars to galaxies of much lower redshift.

The quasar observations by Arp were opposed by the general astronomical community, because they did not agree with the accepted dogma that quasars are billions of light years away. He found it very difficult to get his findings published. Some journals rejected his work, and papers were often held up for years by referees. Finally in 1984, the committee that controls observation time at the Palomar and Mount Wilson Observatories refused to allow Arp to use these facilities.

After being denied research facilities in California, Halton Arp was forced to move to Germany to continue his career, where he joined the Max Planck Institute for Physics and Astrophysics in Munich.

The traumatic experience of Halton Arp was summarized as follows by Fred Hoyle, Geoffrey Burbidge, and Jayant V. Narlikar [20] in their book published in the year 2000, *A Different Approach to Cosmology*:

> "Arp's own colleagues at the Mount Wilson and Palomar Observatories . . . recommended to the directors of the two observatories that his observational program should be stopped, i.e., that he should not be given observing time on the [telescopes in these observatories] to carry on with this program. Despite his protests, the recommendation was implemented, and after his appeals to the trustees of the Carnegie Institution were turned down, he took early retirement and moved to Germany where he now resides, working at the Max-Planck-Institut fur Physik und Astrophysik in Munich. . . . Thus Arp was the subject of one of the most clear cut and successful attempts in modern times to block research which it was felt, correctly, would be revolutionary in its impact if it were to succeed."

Halton Arp was not directly concerned with the Big Bang theory. However, if his observations are correct, the quasar redshift cannot be explained by the Einstein theory, which Big Bang cosmologists use as their "Bible". The general acceptance of the Arp discoveries would cast serious doubts on the validity of that "Bible".

Lack of Scientific Objectivity in Astronomy Today

As we marvel at the astounding advancements of scientific knowledge over the past 400 years, we should recognize that scientific objectivity has been a fundamental requirement in the development of this knowledge. To achieve this objectivity, there must be open scientific debate. Without scientific debate, there can be no real science.

Chapter 2 examined the evidence that has been amassed by paleontologists to develop our understanding of how life evolved on earth. This knowledge has been obtained from a tremendous amount of painstaking work by paleontologists, carefully studying minute details in fossils. To achieve reliable means of interpreting these findings has required extensive scientific debate.

One example of this is the question of the ancestry of birds. Did birds evolve from dinosaurs, or from some other reptile? This relatively minor question has been heatedly debated for years. It is only with such debates that reliable conclusions can be realized.

Astronomy was originally a very objective science. However, since the rapid growth of the Big Bang theory in the 1960's, it has become highly dogmatic. Scientific facts are now forced to fit into a rigid mold.

Astronomical evidence that does not agree with accepted dogma is ignored. We can be sure that many astronomers have learned from the example of Halton Arp, and are careful to find only those results that are consistent with accepted concepts.

Even a superficial assessment will show that there are serious doubts concerning the Big Bang theory. Yet the Big Bang is treated as fact, not theory. Evidence that does not agree with the Big Bang is suppressed. As a result, astronomy has deteriorated into mythology. Conclusions are based on consensus rather than evidence.

Cosmic Microwave Radiation

An important milestone in the development of the Big Bang theory occurred in 1965, when Arno Penzias and Robert Wilson, two physicists working at Bell Laboratories, discovered cosmic microwave radiation. As was shown earlier, this radiation had been predicted by George Gamow. It was claimed to be the relic of optical radiation emitted 300,000 years after the Big Bang.

When cosmic microwave radiation was discovered, it was loudly proclaimed to be proof that the Big Bang must have occurred. However, as will be shown in Appendix C, the Yilmaz cosmology model also predicts cosmic microwave background radiation, and it does so much more accurately than the Big Bang theory.

Cosmic microwave radiation has been claimed to have caused the strong support of the Big Bang theory that occurred after 1965, but this claim is doubtful. The primary cause for that support was clearly economic. There was a tremendous increase in the funding of astrophysical research at that time. Besides, hundreds of scientists found that computer study of the Einstein theory was a rewarding area of research, and the Big Bang theory was the only problem to which these computer studies could be applied. Therefore momentum behind the Big Bang theory exploded with a loud *Bang*.

Chapter 13

Weaknesses of the Einstein Theory

Although the principles of the Einstein theory are very sound, the Einstein gravitational field equation has serious weaknesses. However, these were not apparent during Einstein's lifetime because of the great mathematical complexity of the theory. The Schwartzschild limit was an annoying problem, but Einstein dismissed it as physically irrelevant, since it requires a mass-to-radius ratio that is one-quarter million times greater than the maximum value experienced within our solar system.

In 1939, Oppenheimer and Snyder predicted that a star must collapse to form a *"black-hole"* singularity if its mass-to-radius ratio exceeds the Schwartzschild limit. Einstein flatly rejected this concept, because it drastically violates our laws of physics. He convinced himself that the Einstein gravitational field equation does not predict a singularity. No one could refute Einstein until computers became widely available in the 1960's, about a decade after his death.

The physically impossible Schwartzschild *black-hole* singularity indicates that the Einstein theory has a mathematical weakness. This chapter discusses several other weaknesses of the Einstein theory that have become apparent since Einstein's death. Probably the most serious flaw is that the Einstein gravitational field equation cannot achieve any more than a single body solution.

Does Not Achieve a Two-Body Solution

Professor Carroll O. Alley

Professor Carroll O. Alley of the University of Maryland is one of the very few experts in General Relativity theory who has applied his knowledge to practical applications. Alley has supervised several experiments to test the validity of predictions derived from the Einstein theory. This has included laser measurements with retro-reflectors on the moon that have allowed several distances to the moon to be measured with laser beams to an accuracy of 3 centimeters. Another set of experiments measured the relativistic time delay in an atomic clock carried in an aircraft under several flight profiles. [Y10]

Alley is also intimately involved in applying General Relativity corrections to the Geophysical Positioning System (GPS). The GPS is an array of satellites operated by the U. S. Air Force to provide accurate position coordinates over the world for military and civilian navigation.

Alley became impressed with the Yilmaz theory and is cooperating with Yilmaz. Alley made an important contribution to this issue with his proof that the Einstein theory cannot achieve a two-body solution. Let us examine this astonishing claim.

The Single-Body Schwartzschild Solution

The Schwartzschild analysis of the Einstein theory was only a *single-body solution*. It merely considered the gravitational field of one body, a single star. When this analysis was applied to calculate the relativistic advance of the Mercury orbit, the analysis assumed a test mass in the orbit of Mercury, which had absolutely no effect on the gravitational field.

The orbit of the planet Mercury advances by 1.39 arc seconds for every orbit. This means that the axis of the elliptical Mercury orbit rotates 1.39 arc seconds for every revolution of Mercury, and this rotation is in the direction of the motion of Mercury. This advance of the Mercury orbit was determined by measuring the cumulative advance of the orbit over many years.

Of this 1.39 arc-second advance per orbit of Mercury, 1.29 arc seconds can be calculated using Newton's theory of gravity by considering the forces exerted by other planets on Mercury. The remaining 0.10 arc-second per orbit component of the Mercury advance was calculated using the Schwartzschild solution. The 1.29 arc-second per orbit advance of the Mercury orbit caused by other planets could not be calculated from the Schwartzschild solution, because that solution could only consider the effect of the gravitational field of the sun.

As Mercury rotates around the sun, the gravitational field of Mercury causes the center of the sun to wobble slightly. The Schwartzschild analysis cannot include this effect, because it is only a *single body solution*. It can only account for the gravitational field of a single body, the sun. The gravitational field considered in the analysis is

182 *The Scientific Story of Creation*

not affected by the mass of any planet, including Mercury.

Einstein recognized that his analysis of the Mercury orbit used a *single-body solution*. He and other scientists were content with this limited analysis because it was not practical during Einstein's lifetime to achieve a multi-body solution with the very complicated tensor equation of the Einstein theory. A multi-body solution of the Einstein gravitational field equation would require a non-diagonal metric tensor, which would result in millions of terms in the analysis.

Einstein assumed that his theory could yield a multi-body solution, but the following discussion shows that it cannot. More precisely, *the Einstein theory cannot yield an interactive multi-body solution.*

The Analysis of Professor Alley

Many computer studies using the Einstein theory have appeared to achieve multi-body solutions. However these studies employ artifices that help to make the iterative computer programs converge to solutions. These artifices are inserting results into the solutions that are not actually coming from the Einstein gravitational field equation.

Alley has avoided this problem by considering a simple physical model that can be solved analytically by the Einstein theory. He calculated the gravitational attraction between a pair of infinite slabs of matter separated by a fixed distance. He found that the Einstein theory predicts that there is no gravitational attraction between the two slabs.

This configuration is physically similar to models used in electronics to calculate capacitance. By assuming that the dimensions of the slabs are infinite relative to the separation between the slabs, one can ignore edge effects. This results in a simple theoretical model to which one can apply the Einstein theory analytically. [Y10, Y12]

The analysis shows that the Einstein theory predicts that the gravitational force between the two slabs is zero. This result conflicts with Newton's law of gravitational attraction and with experimental evidence. Chapter 5 described the experiment by Henry Cavendish, which determined the gravitational constant G by measuring the gravitational attraction between lead spheres. Alley's analysis directly conflicts with the Cavendish experiment.

Reason for Failure to Achieve a Two-Body Solution

The following discussion explains in physical terms why the Alley analysis does not predict a gravitational force between the two slabs.

The Einstein, Ricci, and Energy-Momentum Tensors

The space between the two slabs does not contain matter, and so the energy momentum tensor must be zero in that space. Since the Einstein gravitational field equation sets the *Einstein tensor* proportional to the *energy momentum tensor*, the Einstein tensor must be zero in the space between the slabs. If all elements of the Einstein tensor are zero, all elements of the closely related Ricci tensor are also zero.

Gravity is represented as a curvature of space in the Einstein theory, and the Einstein (or Ricci) tensor describes the curvature of space. Since the Einstein curvature tensor is zero in the space between the two slabs, there is no gravitational field in that space to produce gravitational attraction between the two slabs.

The Yilmaz theory does not have this problem, because its gravitational field equation has a stress-energy tensor for the gravitational field. With the Yilmaz theory, the stress-energy tensor for the gravitational field is not zero in the space between the slabs, and so the Einstein curvature tensor is not zero. Consequently the Yilmaz theory predicts gravitational attraction between the two slabs.

In the Einstein theory, the energy momentum tensor is always zero in a vacuum, and so the Einstein curvature tensor and the Ricci curvature tensor are always identically zero in a vacuum.

However, in the Schwartzschild solution, the Einstein curvature tensor is zero in the space outside the star. If there is no curvature in this space, how can the Schwartzschild solution predict that a gravitational force is exerted on the planet Mercury? To understand this we must consider the Riemann tensor.

The Riemann Tensor

The Einstein tensor is a minor modification of the Ricci tensor, and both tensors describe the curvature of space. The Ricci tensor is a reduced version of a very complicated tensor called the Riemann tensor. The Riemann tensor uniquely specifies the curvature of space. If space has no curvature, and hence no gravitational field, all elements of the Riemann tensor are zero. Conversely, if all elements of the Riemann tensor are zero, space has no curvature and hence no gravitational field. The Riemann tensor is not directly used in Einstein theory calculations.

The Riemann tensor has four indices, and hence has 4x4x4x4 elements, or 256 elements. This tensor is denoted in the form $R_{abc}{}^{d}$. The

Ricci tensor is calculated from the Riemann tensor in a process called contraction. The two indices (c) and (d) of the Riemann tensor are set equal to one another, and the Riemann tensor is summed over the four values (0, 1, 2, 3) of these two indices. This summation yields the Ricci tensor R_{ab}, which has 2 indices and hence has 4x4 or 16 elements.

If all elements of the Riemann curvature tensor are zero, then all elements of the Ricci and Einstein curvature tensors are zero. On the other hand, the Ricci and Einstein curvature tensors can be identically zero when the Riemann tensor is not zero. This case occurs in the vacuum of space outside the star in the Schwartzschild solution. Consequently the Schwartzschild solution has a gravitational field in the vacuum of space outside the star.

But the Schwartzschild solution is a very special case. It has a spherically symmetric gravitational field and is only a single-body solution. In this *very special case* the Riemann curvature tensor can be nonzero in the space outside the star, even though all elements of the Ricci curvature tensor are zero. In more general cases, the Riemann tensor must be zero in a region where all elements of the Ricci tensor are zero, because the Ricci tensor is a contraction of the Riemann tensor.

The elements of the Ricci tensor are calculated by adding together appropriate components of the Riemann tensor. How can nonzero elements of the Riemann tensor add to produce elements of the Ricci tensor that are all identically zero throughout a region of space? This can only be achieved in a highly symmetric situation such as occurs in the Schwartzschild solution. In more general cases, the only way that all elements of the Ricci tensor can be zero throughout a region of space is for the Riemann tensor to be zero in that region

In the Alley analysis, all elements of the Ricci tensor are zero throughout the space between the two slabs. Since the geometry does not have spherical symmetry, the Riemann tensor must also be zero, and so there is no curvature (and hence no gravitational field) in the space between the two slabs. Consequently there is no gravitational attraction between the two slabs.

As this discussion has shown, there are fundamental reasons why there is no gravitational attraction between the two slabs in the Alley analysis. ***The Alley analysis proves that the Einstein gravitational field equation cannot achieve an interactive multi-body solution***

Conservation of Matter-Plus-Energy

In order to yield realistic predictions, a relativistic theory must achieve conservation of matter-plus-energy. Since matter can be converted into energy, and vice-versa, it is the sum of matter-plus-energy that must be conserved. This conservation is achieved by placing appropriate constraints on the energy-momentum tensor.

This issue is discussed in Appendix E. Although the Einstein and the Yilmaz theories both have energy-momentum tensors, these tensors are not exactly the same. Yilmaz has proven that the energy momentum tensor for the Yilmaz theory always achieves conservation of matter-plus-energy, but the energy-momentum tensor for the Einstein theory generally does not.

Multiple Solutions from the Einstein Theory

One confusing aspect of the Einstein theory is that it can yield multiple solutions for the same physical model. Constraints that are somewhat arbitrary must be included in the analysis to achieve an answer. This problem is widely recognized by those performing General Relativity studies.

In contrast, the Yilmaz theory has a definite solution and can yield only one answer for a particular physical model. The time-varying Yilmaz theory incorporates general constraints, which assure that this condition is always satisfied.

Scientists who are familiar with the arbitrary adjustable parameters of the Einstein theory may find it difficult to recognize that the Yilmaz theory does not have this property. A prediction made by the Yilmaz theory depends only on the characteristics of the physical model on which it is based. It is not affected by arbitrary assumptions made by the individual who is applying the Yilmaz theory.

The Einstein Theory is Not Rigorous

This discussion has shown that there are fundamental flaws in the Einstein gravitational field equation, which characterizes the Einstein theory. Because of the great mathematical complexity of that equation, its flaws have been obscured.

The success of the Schwartzschild solution has disguised the fact that this is only a single-body solution, which gives reliable results only in a weak gravitational field. The Schwartzschild limit indicates that the Schwartzschild solution does not work in an intense gravitational field, even for a single-body solution. The Alley analysis has proven that the

Einstein gravitational field equation cannot yield an interactive multi-body solution.

The Alley analysis proves that the Einstein gravitational field equation does not provide a rigorous solution to Relativity principles. Confusing results and nonphysical predictions have been derived from the Einstein gravitational field equation because that equation is not mathematically rigorous. The physically impossible singularity predictions of the Big Bang cosmology theories are examples of this.

Variation of Speed of Light with Direction

For tests performed within our solar system, the Einstein and Yilmaz theories yield essentially the same results. Consequently all tests that have verified the Einstein theory are consistent with the Yilmaz theory.

However, there is one very sensitive test, which has been partially implemented, that could distinguish between the Einstein and Yilmaz theories. Yilmaz predicts that the speed of light measured locally in a gravitational field is the same in all directions, but the Einstein theory predicts it is different.

As we saw in Chapter 9, Yilmaz proved that for the speed of light measured locally to be independent of direction, the product ($g_{00}g_{11}$) must be equal to -1, and the elements g_{11}, g_{22}, g_{33} must be equal in rectangular coordinates. These conditions are satisfied by the Yilmaz theory, but are not satisfied by the Einstein theory.

It is very difficult to measure with sufficient accuracy the variation of the speed of light with direction, because the test cannot use two-way transmission of light. As explained by Prof. Carroll O. Alley [Y10], preliminary experiments were performed under his direction to perform this measurement. This involved the transportation of an atomic clock, back and forth between the U.S. Naval Observatory and the NASA Goddard Optical Research Facility, which are 21.5 km apart. The transported clock was used to synchronize atomic clocks in the two locations.

The initial data were very promising. However, government funding was cancelled before measurements could be made with the required accuracy to obtain definitive results. Why was the modest funding for this very important experiment cancelled?

This experiment was discussed by Ivars Peterson [38] in a 1994 *Science News* article: "A New Gravity: Challenging Einstein's general theory of relativity".

Chapter 14

The Yilmaz Cosmology Model

We have seen that the Yilmaz theory of gravity is a refinement of the Einstein General theory of Relativity and has corrected its weaknesses. The Yilmaz theory embodies the principles of the Einstein theory, even though it yields solutions that are different from those obtained from the Einstein gravitational field equation.

Therefore it is reasonable that we should apply the Yilmaz theory to cosmology. We make the simple physical assumptions that the universe has a constant average density of matter that extends to infinity and does not change with time. Although the distribution of galaxies in the universe is clumpy, this assumption of uniform density is adequate as a first-order model of the universe. From this simple model the basic relativistic characteristics of the universe can be found.

Yilmaz presented this cosmology model briefly in the first paper on his gravitational theory, published in the *Physical Review* in 1958. [Y1] Since that time, Prof. Yilmaz has ignored cosmological applications of his theory, because he found that cosmology can be very speculative.

As was shown in Chapter 13, the Yilmaz theory yields unique predictions that depend only on the characteristics of the physical model. Unlike the Einstein theory, the results are not modified by arbitrary decisions made by the individual who is applying the theory. Since the Yilmaz theory has a profound mathematical foundation and its predictions are unique, this cosmology model derived from the Yilmaz gravitational theory deserves our serious consideration.

The author has explored the cosmological implications of the powerful Yilmaz gravitational theory. Analyses were performed to extend the cosmology model that Yilmaz presented in his 1958 paper. These are presented in Appendices C and D of *Believe* [1]. This chapter summarizes the material of Appendix C, which describes the basic characteristics of the cosmology model. The material of Appendix D describes the cosmic microwave radiation predicted by the model, and is presented in Appendix C of this book

Description of the Yilmaz Cosmology Model

Reduction of Speed of Light, Clock Rate, and Spatial Dimensions

We saw in Figs 10-1 and 10-2 of Chapter 10 that the gravitational field of a star causes the speed of light to decrease, it causes a spatial dimension to contract, and it causes a clock to run slower. The Yilmaz Cosmology Model predicts that the gravitational field produced by matter in the universe should cause similar effects as we look out into space. The results are shown in Fig. 14-1. The solid curve shows how the speed of light should decrease with distance, and the dashed curve shows how a clock rate and a spatial dimension should decrease with distance.

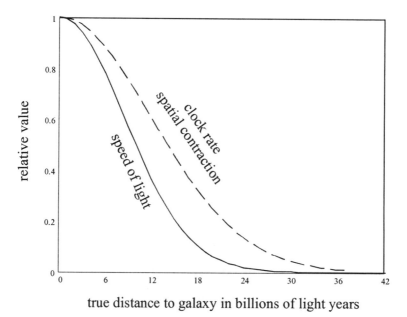

Figure 14-1: Apparent speed of light (solid), clock rate (dashed), and spatial contraction (dashed), versus distance to a galaxy

These plots for the Yilmaz cosmology model assume that the Hubble constant H_0 is 25 km/sec per million light years, which corresponds to an apparent age of the universe T_0 of 12 billion years. As was explained earlier, recent studies show that the Hubble constant is probably close to 20 km/sec per million light years, which corresponds to an apparent universe age of 15 billion years. Nevertheless, there is still appreciable uncertainty in the Hubble constant, and so this book continues to use for the Yilmaz cosmology model a Hubble constant of 25 km/sec per million light years and an apparent universe age of 12 billion years.

This analysis is based on the parameter r_0, which is the distance that light travels during the apparent age T_0. For the Yilmaz cosmology model, we assume that r_0 is 12 billion light years.

The plot of spatial contraction given in Fig 14-1 shows that a dimension appears to contract with distance, and so the apparent distance to a galaxy is decreased. Figure 14-2 gives a plot of the apparent distance to a galaxy versus the true distance. Because of the strong contraction of spatial dimensions at great distances, the maximum apparent distance to any galaxy is finite, even though the model assumes that the true galaxy distance extends to infinity.

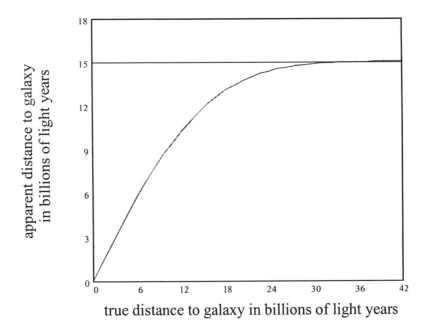

Figure 14-2: Apparent distance to galaxy verses true distance

The maximum apparent distance to a galaxy is equal to $\sqrt{[\pi/2]}\, r_0$. We assume r_0 to be 12 billion light years, and for this value of r_0 the maximum apparent distance is 15.040 billion light years, or approximately 15 billion light years.

There is a very high contraction of the dimension of a galaxy that is close to the apparent limit of 15 billion light years. Consequently the apparent density of matter becomes extremely high as this limit at 15 billion light years is approached. This effect is shown in Fig. 14-3.

Figure 14-3 shows how the density of the universe appears to vary with apparent distance from the earth. The darker the picture, the greater is the apparent mass density. The circles are contours of constant apparent mass density, and correspond to relative mass density values of 1.5, 3, 10, 30, 100, and 1000.

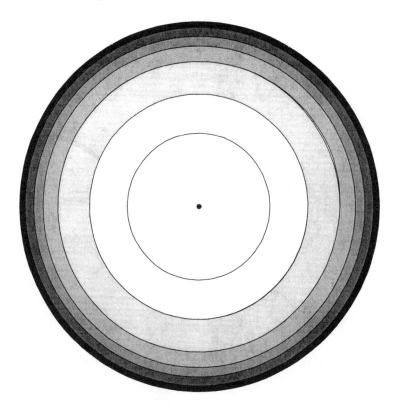

Figure 14-3: Apparent relative mass density of universe seen from earth; boundaries at density values of 1.5, 3, 10, 30, 100, 1000; minimum and maximum radii at 6 and 15 billion light years.

The apparent distance of the inner circle is about 6 billion light years, and that of the circumference is about 15 billion light years. The circumference theoretically corresponds to an infinite density of matter. However, light cannot reach us from galaxies beyond 45 billion light years of true distance. Hence the apparent mass density is not infinite at the circumference, although it can be very large.

We should realize that the "apparent" effects shown in Figs. 14-1 to 14-3 are real effects. As was explained in our discussion of Special Relativity, *Reality is Relative*. There is no such thing as *absolute reality*. Any observation that we make from earth will display the "apparent" characteristics of the universe that we have examined.

Nevertheless we can still consider a "true" picture of the universe where galaxies are at a "true" distance from the earth. This "true" picture gives us a simple absolute model for understanding the universe. An interval of "true distance" represents an interval of "proper distance" that would be measured in proper coordinates situated at that location.

Figure 14-2 shows that a distant galaxy appears to be closer than it actually is. Nevertheless, the apparent time for the light to reach us from the galaxy is not reduced, because the apparent speed of light and the apparent clock rate also decrease with distance. Combining the spatial contraction, speed of light, and clock rate plotted in Fig. 14-1 shows that the apparent time for light to travel between two distant points is the same as if there were no relativistic effects.

For example, consider a distance where the spatial contraction is ½. The relative speed of light is ¼ and the clock rate is ½. With distance reduced to ½, light should take twice as long to travel between two points if the speed of light is ¼. However, the apparent clock rate is cut in half, and so the apparent time for light to travel between the two points is the same as if there were no relativistic effects.

The Hubble Expansion of the Universe

In General Relativity theory, the *geodesic equations* are used to calculate the trajectory of a planet or any other body, or of a particle, or even of a light photon. To verify the Einstein theory, the geodesic equations were applied to the Schwartzschild solution to calculate the bending of a light ray when it passes close to the sun, and the relativistic advance of the orbit of Mercury.

When the geodesic equations are applied to the Yilmaz cosmology model, they show that a distant galaxy must recede at a velocity approximately proportional to its distance. In other words, the universe must expand just as Hubble observed. The results of this analysis are shown in Fig. 14-4. The solid curve shows the ratio of apparent receding velocity V_{ap} of a galaxy relative to the apparent speed of light c_{ap}.

The dashed line in Fig. 14-4 shows for comparison the ideal Hubble law, in which galaxy velocity is exactly proportional to distance. According to the Hubble law, the velocity of a galaxy would reach the speed of light at a distance r_0 of 12 billion light years, and would exceed the speed of light at greater distances.

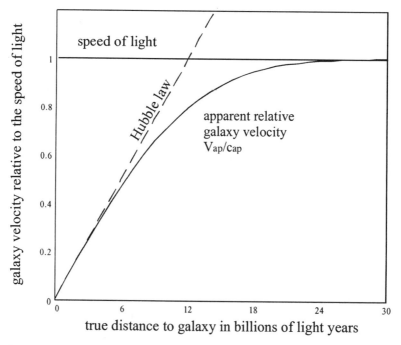

Figure 14-4: Apparent galaxy velocity relative to apparent speed of light, compared with Hubble law.

The solid curve shows that the receding velocity of a galaxy approximates the ideal Hubble law plot out to about 5 billion light years. Thus the universe should expand approximately in accordance with the Hubble law within 5 billion light years. At much greater distances the apparent galaxy velocity gradually approaches the apparent speed of light, but never exactly reaches it.

The expansion rate that is predicted by the Yilmaz cosmology model varies with the mass density of the universe. The greater the density of matter, the faster the universe should appear to expand. For our assumed Hubble constant, 25 km/sec per million light years, the average density of matter is equivalent to 7.5 hydrogen atoms per cubic meter. A Hubble constant of 20 km/sec per million light years would result in an average density of matter of 4.8 hydrogen atoms per cubic meter.

The solid curve in Fig. 14-4 represents the ratio V_{ap}/c_{ap}, which is the apparent receding velocity divided by the apparent speed of light. It is this apparent velocity ratio that astronomers calculate from the Doppler wavelength shift of galaxy spectra. They assume that they are measuring the true galaxy velocity relative to the true speed of light, but they are only measuring the apparent value of this ratio.

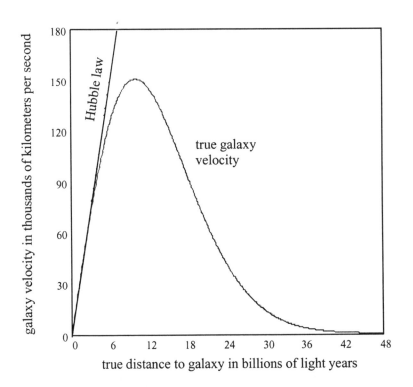

Figure 14-5: True galaxy velocity expressed in thousands of kilometers per second compared with Hubble law.

To see how the universe is actually expanding, we need the ratio V/c, which is the true galaxy velocity V divided by the speed of light c measured on earth. The spatial contraction plot in Fig. 14-1 is equal to the velocity ratio V_{ap}/V, and Fig. 14-1 also gives the relative speed of light c_{ap}/c. The following plots are combined to obtain the V/c ratio: the plot of V_{ap}/c_{ap} in Fig. 14-4, the speed of light plot of c_{ap}/c in Fig. 14-1, and the spatial contraction plot in Fig. 14-1, which is equal to V_{ap}/V. The resultant plot of V/c is shown in Fig. 14-5.

In Fig. 14-5, the speed of light c was set equal to 300,000 km/sec to give the true receding velocity V of a galaxy in absolute terms. This shows that the true velocity V of a galaxy reaches a maximum value of 150,000 km/sec at a true distance r of 10 billion light years. This maximum value is half the speed of light c measured on earth.

At greater distances the true galaxy velocity decreases, and becomes very small at large true distances. For example, at a true distance of 36 billion light years, the true galaxy velocity is only 3000 km/sec, which is one percent of the normal speed of light, and at a true distance of 48 billion light years, the true galaxy velocity is only 100 km/sec.

Thus Fig. 14-5 shows that over very large distances the universe does not expand. It demonstrates that the Hubble expansion is a local relativistic distortion of space, not a general expansion of the universe.

How Can Gravity Make the Universe Expand?

The expansion of the universe displayed in Figs. 14-4 and 14-5 is a consequence of the Yilmaz gravitational theory. This indicates that the Hubble expansion of the universe is caused by gravity. "How can this be?", you ask, "How can gravity, which always causes masses to attract one another, force the universe to expand?"

Part of the answer is that the Hubble expansion is a local effect. Over very large distances the universe does not expand. Yet we are still faced with the question, "How can gravity make the universe expand locally?"

The rigorous mathematical answer is that the geodesic equations, which characterize the effect of gravity, show that the universe must expand. This issue in discussed in Appendix C of *Believe* [1] and is proven mathematically in Chapter 4 of the website *Addendum* [3]. Nevertheless we still would like an intuitive answer that makes sense physically.

Let us consider the following intuitive explanation. As was shown in Fig. 14-3, the whole universe appears to be compressed within a sphere having a radius of about 15 billion light years. The figure shows that the apparent density of matter is extremely high near the periphery of this sphere. We can assume that this outer shell of high-density matter is exerting gravitational force on matter that is inside the sphere. This gravitational force could be pulling matter toward the periphery, thereby producing the Hubble expansion. Thus gravitational attraction could cause a local expansion of the universe.

However, this simple intuitive explanation is not consistent with gravitational force as described by Newton. Suppose that the universe is modeled as a thin spherical shell, in which matter is evenly distributed over the shell. If Newton's law of gravity is applied to a mass element placed inside the shell, the gravitational forces on the mass are exactly cancelled. There is no net force attracting the mass element toward the spherical shell.

On the other hand, Newton's laws are only approximated in this relativistic model. If gravitational forces inside the shell do not cancel exactly, there can be a net gravitational force attracting matter toward the mass of the shell. This interpretation appears to explain how the attractive effect of gravity can cause the universe to expand locally.

This gives a simple intuitive explanation of how gravity could cause the universe to expand. On the other hand, as was stated earlier, the rigorous answer is that the universe expands locally because the geodesic equations show that it must.

Creation of Matter

The Yilmaz cosmology model assumes that the average density of matter does not change with time. In order to satisfy this requirement as the universe expands locally, the cosmology model requires that matter must be created to compensate for the expansion. Since matter and energy are equivalent, the creation of matter could be achieved by converting energy into matter.

The Yilmaz gravitational theory requires that matter-plus-energy must be conserved. Therefore the Yilmaz cosmology model implies that energy is radiated across the universe to create matter that compensates for the Hubble expansion. This energy would be derived from matter in other parts of the universe.

The rate of creation of matter that is required to compensate for the Hubble expansion is one hydrogen atom created per cubic meter every 500 million years. The rate of conversion of energy into mass to achieve this creation of matter is equal to 10 microwatts of power continually converted into matter within a volume the size of the earth.

Over very large distances the universe does not expand. Hence the total amount of matter and energy in the universe stays constant.

Cosmic Microwave Background Radiation

Proponents of the Big Bang theory strongly acclaim the discovery of cosmic microwave background radiation as a milestone in validating the theory. In 1965 this radiation was first detected in a sensitive communication antenna at Bell Laboratories. Much more accurate measurements were obtained from the Cosmic Background Explorer (COBE) satellite in 1989. These experiments show that microwave background radiation is emanating uniformly from all directions, and has the spectrum and intensity that would be emitted from an ideal blackbody at a temperature of 2.73 degrees Kelvin.

Cosmic microwave background radiation was predicted by Big Bang theorists, who claimed it to be the cooled relic of radiation emitted from hot plasma 300,000 years after the Big Bang. However estimates of the blackbody temperature by Big Bang theorists varied from 5 °K to 30 °K, and so the cosmic radiation was only predicted in a qualitative sense.

The Yilmaz cosmology model also predicts cosmic microwave radiation. Figure 14-4 shows that at very large distances the apparent galaxy velocity is very close to the apparent speed of light, and so the light radiated from distant galaxies should be Doppler shifted to very low frequencies. *Believe* [1] analyzes this radiation in Appendix D, and this analysis is summarized in Appendix C of this book. The predicted cosmic radiation is equivalent (in both spectrum and intensity) to the emission from an ideal blackbody at a temperature between 2.1 °K and 3.4 °K. This is highly consistent with the 2.73 °K blackbody temperature measured by the COBE satellite.

Uniqueness of Cosmology Model Predictions

The Yilmaz gravitational theory gives a rigorous and unique relativistic specification of the effects of gravity. The Yilmaz cosmology model applies this gravitational theory to a simple physical model of the universe. Since the Yilmaz theory is rigorous, and its predictions are unique, the description of the universe that has resulted from the Yilmaz cosmology model should be taken seriously.

For scientists accustomed to the Einstein theory, these principles may be difficult to understand, because the Einstein theory yields multiple, contradictory solutions. In contrast, the predictions of the Yilmaz theory are unique, and so are not affected by arbitrary assumptions made by the individual who is applying the Yilmaz theory.

Chapter 15

A Believable Picture of the Universe

Viable Models of Our Universe

Let us return to the question, "How was Our Universe Created?" In the 1950's there were two competing answers to this question, the Gamow Big Bang theory and the Steady-State Universe theory.

Whether or not the Gamow Big Bang theory is correct, it at least is physically realistic. Gamow postulated that the universe began as an extremely compact body with the density of a neutron star. With the Gamow postulate, the initial Big Bang universe would have just fit inside the orbit of Mars.

Modern Big Bang cosmologists have concluded that our universe began as a *singularity*. Astrophysicist P. James E. Peebles, acclaimed the *"father of modern cosmology"* by *Scientific American*, estimated that the initial universe was "concentrated within a region smaller than a dime" at the instant of the Big Bang. This singularity concept was derived from computer studies of the Einstein General theory of Relativity. ***Nevertheless, Einstein solidly rejected the singularity predictions that had been derived from his theory. Einstein insisted that, "singularities do not exist in physical reality".***

According to our normal laws of physics, a neutron star has the greatest density of matter that is physically possible. Yet modern Big Bang cosmologists maintain that an unknown physical process allowed matter at neutron-star density to collapse from the *orbit of Mars* down to the *size of a dime*. This prediction is so far removed from physical reality it demonstrates that the modern Big Bang cosmology theories are devoid of scientific validity.

This leaves us with the Gamow Big Bang theory and the Steady-

State Universe theory as primary hypotheses for how our universe was created. The original Steady-State Universe theory was abandoned by its sponsors primarily because it did not have a good answer for cosmic microwave background radiation. However, as was shown in Chapter 14, the Yilmaz theory yields a universe model that agrees with the basic principles of the Steady-State Universe theory but does not have the weaknesses of the original theory.

The only physical assumptions of the Yilmaz cosmology model are that the universe has a constant average density of matter that extends to infinity and does not change with time. Although galaxies are not distributed uniformly in our universe, this assumption of a constant average density of matter should be adequate to achieve a first-order relativistic model of our universe.

This chapter examines the picture of the universe that is predicted by the Yilmaz cosmology model. Is this picture correct? *Since the only viable alternative to the Yilmaz cosmology model is the Gamow Big Bang theory, this picture deserves our serious consideration. The modern Big Bang theories, with their singularity postulates, must be discarded as mathematical fantasies.*

The Implications of the Yilmaz Cosmology Model

Let us explore the picture of the universe that has been predicted by the Yilmaz cosmology model. We start by returning to the principles of the Steady-State Universe theory, which assumes that the age of the universe is infinite. The theory assumes that the universe has always appeared approximately like we see it today, and looks roughly the same from every point of the universe. As the universe expands, matter is being created to compensate for the expansion. This matter forms new stars and galaxies, and so the universe continually changes in detail.

Our picture of the universe accepts these basic principles of the Steady-State Universe theory. However, the Yilmaz cosmology model does not suffer from the weaknesses that caused the original Steady-State Universe theory to be abandoned.

We postulate that diffuse matter is continually being created throughout the universe. We assume that matter is created one atom at a time as a hydrogen atom, which has a single proton and an electron. This creation of matter is achieved by the conversion of energy into matter, and so is consistent with our laws of physics. The energy is derived from radiation emitted by stars throughout the universe.

How is this conversion of energy into matter achieved? Although we do not know this answer, this postulate does not violate any law of physics and is consistent with observational evidence.

The required rate of creation of matter is equivalent to one hydrogen atom created within a volume of one cubic kilometer every six months. One hydrogen atom would be created every half-million years within a cube that is 10 meters on a side, which has the volume of a medium-size house. This creation of matter would be achieved by the conversion of energy into matter. The rate of energy utilization is equivalent to the continual conversion of 10 microwatts of power into matter within a volume the size of the earth.

Because of gravitational forces, the diffuse matter that is continually created throughout the universe eventually congregates into enormous clouds of hydrogen. A cloud of hydrogen forms a galaxy, and bits of the galaxy coalesce to form individual stars.

Matter Derived from Gravitational Waves

We postulate that energy radiated from stars in the form of electromagnetic waves (light, heat, X-rays, etc.) is transmitted across the universe and is converted into diffuse matter in space. This matter in turn coalesces to form new stars and galaxies, and the process continues indefinitely.

However, this postulate alone could not result in a Steady-State Universe that lasts indefinitely. The universe would gradually deteriorate into black dwarf and neutron stars, and so would eventually die. There must be a process that converts the mass of these dead stars into energy.

The Einstein and Yilmaz theories both predict gravitational waves. As explained in *Scientific American*, April 2002 (pp. 62-71), scientists are performing elaborate experiments that are attempting to measure gravitational waves. We postulate that the mass of a black dwarf or a neutron star is converted into gravitational waves that are radiated. By this means, the matter within these dead stars would be gradually dissipated as radiated energy. The extreme densities of black dwarf and neutron stars may accelerate the generation of gravitational waves.

Therefore we postulate that energy is radiated from stars in the form of electromagnetic waves and gravitational waves, and this energy is converted into diffuse matter throughout the universe. An interesting possibility is that the creation of diffuse matter might be caused by an interaction between electromagnetic waves and gravitational waves.

We do not know how the energy radiated from stars might be transformed into matter. Nevertheless, this postulate is consistent with our laws of physics. In contrast, the Big Bang theory is based on the physically impossible postulate that all of the matter and energy of the universe was created *out of nothing*, at the instant of the Big Bang.

Our cosmology model envisions a universe of infinite age. New stars in our universe are steadily being formed, as energy radiated from older stars is converted into diffuse matter in space, which condenses to form new stars and galaxies. Because of the continual creation of matter, the universe is always changing, and so our universe stays eternally young even though it is infinitely old.

The Local Expansion of the Universe

A strange and fundamental aspect of the Yilmaz cosmology model is that the expansion of the universe is a local effect. Over very large distances the universe does not expand. The Yilmaz cosmology model predicts that the universe expands locally about every point, yet the over all size of the universe remains constant. How is it possible for the universe to expand everywhere yet not grow any bigger?

This relativistic concept evolved rigorously from the mathematics of the Yilmaz gravitational theory. It is not a product of speculation, and it surprised the author when it was discovered. Although this concept may contradict our usual physical notions, it is no more counter-intuitive than the relativistic principles of Special Relativity, explained in Chapter 7. The principle that two observers moving at different velocity must measure the same speed of light strongly conflicts with our intuition.

A fundamental problem with the original Steady-State Universe theory is that its postulated universe continually grows larger, and has been doing so for an infinite period of time. In contrast, the Yilmaz cosmology model predicts a Steady-State Universe that has a constant size, and within this universe matter-plus-energy is conserved. Although the relativistic principles of this model may seem unbelievable, they provide a coherent means of explaining profound cosmological issues.

The Second Law of Thermodynamics

Another problem with the Yilmaz cosmology model is that its prediction of an infinitely old universe seems to conflict with the principles of thermodynamics. Scientists often use the *Second Law of Thermodynamics* to prove that our universe must eventually run out of available energy and so will die. This law of physics seems to prohibit the possibility that the universe can have an infinite age.

The United States Patent Office does not prohibit a patent on a perpetual motion machine. It merely requires that the inventor submit a working model of the invention before it can be patented. There are two types of perpetual motion machines, depending on which law of thermodynamics is violated. These two laws of thermodynamics can be expressed in simple terms as follows:

> ***First Law:*** *Energy can neither be created nor destroyed. Since matter and energy can be converted into one another, this is expressed in a more general form as: The sum of matter-plus-energy can neither be created nor destroyed.*

> ***Second Law:*** *The availability of energy must continually decrease.*

The Second Law is rigorously expressed in terms of a concept called *entropy*, which roughly means the *degree of disorder*. The Second Law states that *entropy* (the degree of disorder) must continually increase. However, the above definition gives a simpler interpretation of the Second Law.

The *Second Law* can be explained as follow. Consider a train locomotive driven by a steam engine. Water in the boiler is heated to generate steam, which pushes against the pistons to drive the wheels of the locomotive. It can be shown that the energy that can be extracted from the heated steam depends on the difference in temperature between the steam in the boiler and the temperature (100 °C) at which the exhausted steam condenses to water.

The gasoline engine in an automobile achieves much greater efficiency. Its energy output depends on the difference of temperature between the exploding gasoline in the cylinders and the temperature of the air into which the gasses are exhausted. Gasoline engines are run as hot as practical in order to increase their efficiency. Diesel engines achieve greater efficiency than regular gasoline engines because they operate at higher temperatures.

A motor can theoretically run from any source of heat, provided that there is a sink at a lower temperature into which the heat can be discharged. Useful energy cannot be derived from a heat source without a temperature differential.

With the passing of time, our world becomes more uniform. Temperature differentials decrease, and so energy becomes less available. Our earth is replenished by energy radiated from our sun, which is generated by nuclear fusion occurring within the sun. Eventually the sun will run out of nuclear fuel, and will fade to become a dark body. After that occurs, the available energy on the earth will gradually drop to zero.

Thus, in accordance with the *Second Law of Thermodynamics*, the availability of energy within our universe seems to be continually decreasing. Although matter-plus-energy remains constant within our universe, the availability of energy should theoretically decrease until the universe dies. How can we justify a universe that is infinitely old?

The Big Bang theory postulates that an enormous amount of energy and matter was created suddenly *out of nothing* at the instant of the Big Bang. This theory violates both laws of thermodynamics at the moment of creation.

The original Steady-State Universe theory postulates that matter is being created very slowly *out of nothing* to form diffuse matter in space. Like the Big Bang theory, this theory violates both laws of thermodynamics, but does so continuously in infinitesimal amounts.

The Yilmaz cosmology model requires the conservation of matter-plus-energy, and so it satisfies the First Law of Thermodynamics. But what about the Second Law?

Our model postulates that energy radiated from stars forms diffuse matter in space, which coalesces to create new stars. These new stars radiate energy, which forms new diffuse matter, and the process continues indefinitely. Matter-plus-energy is being conserved, as the *First Law of Thermodynamics* requires. However, the availability of energy does not decrease with time, and so the *Second Law of Thermodynamics* is violated. How do we justify this?

This contradiction is explained by the principle of *Relativity*, which allows the universe as a whole to violate the *Second Law of Thermodynamics*. Energy is radiated from distant regions of the universe to generate the matter that compensates for the Hubble expansion. Since *Reality is Relative*, relativistic effects can allow the *Second Law* to be violated by the universe as a whole, even though this law is satisfied locally.

In Chapter 1 we discussed the cosmological theory of Paul Marmet, which postulates that the Hubble redshift is an apparent effect, caused by photon collisions with hydrogen atoms. Marmet postulates that the Hubble redshift does not represent an actual expansion of the universe.

The Marmet cosmology theory seems doubtful because it requires a density of matter that far exceeds astronomical evidence. However, a more basic limitation of the Marmet theory is that it does not account for the Second Law of Thermodynamics. If the Marmet cosmology theory were correct, the Second Law would eventually cause the universe to run out of available energy and die. If the universe must eventually die, it must have had a beginning. What was that beginning?

Since the Hubble expansion was discovered in 1929, it has been regarded as a perplexing enigma that requires an explanation. *Our investigation has shown that the Hubble expansion is not an enigma; rather, it is an essential feature if a universe is to endure forever.* The Hubble expansion (with the associated conversion of energy into matter) allows the universe to change continuously and thereby to stay eternally young although it is infinitely old. Relativistic processes compensate for the *Second Law of Thermodynamics* to keep the universe from running out of available energy.

The Contents of the Universe

The "Observable" Yilmaz Universe

How large is our universe, and how much matter does it contain? As shown in Appendix C, our analysis indicates that cosmic microwave radiation is the Doppler-shifted effect of optical radiation emitted from galaxies at a true distance of 45 billion light years. Since this analysis agrees closely with measured COBE data, we can conclude that the universe should extend uniformly to a true distance of at least 45 billion light years.

Light radiated from beyond 45 billion light years cannot reach us, because it is absorbed by diffuse matter in space. Therefore we can regard 45 billion years to be the radius of the observable universe according to the Yilmaz cosmology model. Galaxies at this 45 billion light-year limit of our universe cannot be observed individually. The radiation from these galaxies is smeared to form the cosmic microwave radiation.

In Chapter 4, Table 4-2 (p. 67) gives an estimate of the stars and galaxies within an observable Big Bang universe with a radius of 15 billion light years. Since the observable universe according to the Yilmaz cosmology model has a radius of 45 billion light years, the ratio of the volumes of these two observable universes is $(45/15)^3$, which is a factor of 27. This shows that the observable universe according to the Yilmaz cosmology model should have 27 times as many galaxies as the observable Big Bang universe.

Accordingly, we should multiply the data in Table 4-2 by 27 to obtain our estimates for the number of stars and galaxies in the observable universe specified by the Yilmaz cosmology model. This gives about 380 billion galaxies that contain luminous stars equivalent to 1.5 billion times one trillion suns.

Appendix C shows that the Yilmaz cosmology model should have 47×10^{21} times the mass of our sun within a sphere that is 15 billion light years in radius. Multiplying this by 27 gives 1.27×10^{24} times the mass of our sun for the total matter within the observable Yilmaz universe. In words this is 1.27 trillion times the mass of one trillion suns.

Is the Universe Infinite?

We may never be able to determine whether the size of the universe is finite or infinite. However it seems more comforting to believe that it is finite. We might assume that the universe folds back onto itself, and so is finite even though it has no boundary.

At the limit of the observable universe, the apparent receding velocity of a galaxy is very close to the apparent speed of light. However, the apparent speed of light at that distance is only 234 meters per second. The true galaxy velocity is 280 km/sec, which is about 1000 times greater.

At a true distance of 60 billion light years, which is beyond the observable limit of 45 billion light years, the true receding velocity of a galaxy should be a mere 1 km/sec, which is only 1/30 of the velocity of the earth around the sun.

Let us assume that the universe folds back onto itself and has an effective radius of about 60 billion light years. This means that if one could travel 60 billion light years in any direction, one would reach the same point. With this postulate, the volume of the total universe would be $(60/45)^3$ times the volume of the observable universe, which is somewhat greater than a factor of 2.

Therefore we double our estimates for the observable universe to obtain the contents for the total universe. We estimate that our total universe has 750 billion galaxies and contains luminous stars equivalent to 3 billion times one trillion suns. The total estimated mass of our universe is 2.5 trillion times the mass of one trillion suns. Remember that one trillion is one million times one million.

Yet our sun has one million times the mass of our earth, which is so infinite to our senses that most people thought the earth was flat until the days of Columbus. From the point of view of a mere mortal, our universe might just as well be infinite.

Religious and Philosophical Implications of Our Picture of the Universe

The Biblical Story of Creation

The Big Bang instant of creation has often been related to the story of Creation in the Bible. Many have claimed that the Big Bang theory strongly reinforces this Biblical story. Nevertheless, we can find a solid basis for supporting the Biblical account of Creation without assuming that our whole universe began with a Big Bang 15 billion years ago.

We know that our sun was created 5 billion years ago. The earth, which defines our world, was created as a molten mass 400 million years later. The earth cooled and solid land was created. Our oceans were created from condensed steam and water from meteorites. The photosynthesis from life containing chlorophyll created the oxygen in our atmosphere, which allows us to breathe. The scientific story of the creation of our sun and our earth and the life on earth is all that one needs to support the principles of the Biblical story of Creation.

This book is not claiming that the story of Creation in the Bible is a scientific account of creation. On the other hand, it is difficult to understand why the Biblical description of Creation has been interpreted as support for the Big Bang theory. The Bible begins with:

"In the beginning, God created the heavens and the earth. The earth was without form and void, and darkness was upon the face of the deep" . . . *"And God said, 'Let there be light', and there was light."*

This is followed by a description of the development of our earth. Since the sun was created before the earth, the above quotation must be referring to the formation of our sun if the account is to have any scientific validity.

One could interpret this quotation as an explanation of the process that created our sun with its solar system. A cloud of hydrogen condensed until nuclear fusion was ignited to produce our sun, "and there was Light". Our earth was initially "without form and void", a collection of dust and gas particles that coalesced to produce our earth.

Thus the Creation story of the Bible is consistent with the formation, 5 billion years ago, of our sun and its solar system, including our earth.

Whether or not this creation of our sun and earth was preceded by the Big Bang creation of our whole universe, 10 billion years earlier, is not even hinted at in the Bible. *The Biblical story of Creation neither agrees nor disagrees with the Big Bang theory.*

Our Picture of the Universe

Let us return to the scientific evidence. The Yilmaz cosmology model predicts that the universe as a whole has always existed, but individual stars and galaxies are continually being created. Matter is created to compensate for the local expansion of the universe. Since this matter is derived from energy radiated from distant stars, the total mass and energy of the universe stays constant.

How much matter is in the universe? We have estimated it to be 2.5 trillion times the mass of one trillion suns, which is twice the estimated mass within the observable universe. Alternatively, we could assume that our universe is infinite. The universe has always had the same amount of matter and energy, and the total does not change with time.

Thus our picture portrays an enormous universe having a constant size. We may never be able to determine whether our universe is finite or infinite. The universe has always existed, appearing approximately like we see it today, and will always remain that way. Yet the universe is continually changing because of the conversion of energy into matter. This creation of matter is what keeps the universe eternally young although it is infinitely old.

This is our picture of the universe. It is not a product of speculation. It evolved quantitatively from the Yilmaz theory of gravity, which has a rigorous mathematical foundation and is a direct refinement of the Einstein General theory of Relativity. I personally find this universe picture to be warmly consistent with my religious beliefs. Those who reject religion should also find it to be philosophically satisfying.

Our picture envisions a universe of infinite age. The universe has always existed and will always continue to exist. Five billion years ago, nuclear fusion was ignited in a collapsing cloud of hydrogen, and our sun was born. *"Let there be Light, and there was Light"*. That event was the moment of Creation as far as mankind is concerned. Our earth was created 400 million years later. Over billions of years life developed on earth to produce the world that we know today.

Chapter 16

The Genius of Albert Einstein

Albert Einstein has been widely acclaimed to be the greatest scientist of the 20th century. Although most people support this conclusion, very few understand what Einstein actually did. An important goal of this book is to explain Einstein's achievements in a simple manner.

Einstein's Discovery of Relativity

The Basic Relativity Principle

The essence of Albert Einstein's genius was his ability to derive broad and radical scientific principles from physical evidence. In attempting to explain the constancy of the speed of light, the great physicist Hendrik Lorentz calculated the manner in which distance and time measurements must change when a body moves with respect to the aether.

It took Einstein to recognize the true concept. We can measure a relative velocity, but absolute velocity has no meaning. Consequently, the *aether*, which theoretically provides an absolute reference for specifying the speed of light, cannot exist.

Since there is no such thing as absolute velocity, Einstein recognized that the speed of light cannot change from observer to observer. Two observers must measure exactly the same value for the speed of light regardless of the relative velocity between them.

To satisfy constancy of the speed of light, Einstein concluded that distance and time intervals must be relative. They must vary with the velocity of the observer that is measuring them. Dimensions, clock rates, and time differences depend on the velocity of the observer. These measurements are relative. Reality is relative.

Two events that occur at the same time in the reference frame of one observer can occur at different times to another observer moving at a different velocity. Consequently we cannot separate time and spatial measurements. They must be combined into a four-dimensional space-time specification.

From these relativistic principles, Einstein derived several profound physical conclusions. One was that mass can be converted into energy. This concept showed how the energy of our sun is produced and eventually led to the harnessing of atomic energy in nuclear power plants and in the atomic and hydrogen bombs. It is remarkable that the abstract principles of Relativity could explain the generation of the enormous energy that is radiated by our sun.

Generalizing the Relativity Principle

After developing his Relativity principle, Einstein soon realized that his basic theory had a serious weakness. It applies only when the observers are moving at constant velocity. His basic theory does not hold when the velocity of an observer changes, which means that the observer is accelerating. Einstein demonstrated that acceleration and gravity are equivalent, and so his basic theory does not hold exactly in a gravitational field. To obtain a rigorous theory of Relativity, Einstein had to generalize his theory to account for the effects of acceleration and gravity.

To achieve this generalization, Einstein adopted a daring approach. Relativity involves the transformation of measurements from one coordinate system to another. To satisfy this in a rigorous manner, Einstein employed the abstract mathematical theory that had been developed by Ricci and Riemann, now called tensor analysis, which expresses calculus in terms of curved space. This *absolute differential calculus* is powerful and rigorous, but is a very complicated mathematical tool.

Tensor analysis has a general formula for translating a tensor from one coordinate system to another. Therefore if a physical law is specified in terms of tensors, it will be the same in all coordinate systems, and so the *Principle of Covariance* is satisfied.

But how can the physical principles of Relativity, including the effects of gravity, be expressed in terms of the abstract mathematics of tensor analysis? Einstein characterized gravity to be a curvature of space. Since tensor analysis deals with curved space, it provided a mathematical means of handling the effects of gravity.

Einstein expressed tensor analysis in four-dimensional space-time coordinates. Since tensor analysis is very general, it was readily applied in four dimensions. Einstein designed his theory so that it reduces to the equations of Special Relativity when the gravitational field is zero. He also required his equations to approximate Newton's theory of gravity in a weak gravitational field.

Einstein postulated that the effects of acceleration and gravity are equivalent. With this principle he was able to calculate the approximate relativistic effects produced by a gravitational field. He constrained his theory to satisfy these approximations.

These are some of the many issues that Einstein considered to develop his General theory of Relativity. It took him 11 years of intensive research before he achieved his final theory, which he presented in 1916. His theory was specified in terms of his gravitational field equation.

Einstein took a bold approach to solve an extremely difficult theoretical problem. However, this book has demonstrated that Einstein did not achieve a rigorous solution to the principles of Relativity. The great mathematical complexity of his gravitational field equation obscured this weakness during his lifetime.

Nevertheless, this limitation of the Einstein theory should not diminish our great respect for Einstein's genius. The principles that Einstein established were very sound. Yilmaz has refined the Einstein theory and has thereby corrected its weaknesses. This refinement has proven that the bold Einstein approach was basically correct.

The Yilmaz Refinement of the Einstein Theory

As we have seen, Yilmaz studied the steps that Einstein took in developing his General theory of Relativity. Yilmaz had the advantage of approaching these issues with a fresh mind. He studied a calculation that Einstein had implemented in an approximate manner, and realized that he could solve it exactly. To this problem Yilmaz applied Special Relativity principles that Einstein had developed.

From this analysis Yilmaz derived an exact formula for the wavelength increase produced by a gravitational field, which in turn showed the effect of gravity on clock rate. Yilmaz calculated from this the metric tensor element (g_{00}) associated with time.

Yilmaz postulated that the speed of light should be independent of direction when measured locally in a gravitational field. With this postulate he was able to calculate the other elements of the metric tensor.

In this manner Yilmaz achieved an exact and rigorous solution to the principles of Relativity that Einstein had established.

These discoveries formed the basis for the Yilmaz theory of gravity. Since his initial discovery, Yilmaz has devoted his life to refining his theory. His basic theory was only a static solution. After 15 years of intensive research, Yilmaz developed his general time-varying theory. Later he proved that his theory is consistent with quantum mechanics, a property that the Einstein theory lacks.

Yilmaz has developed a profound theory that has enormous potential. Since this new theory has applied the principles that Einstein established in his General theory of Relativity, it has demonstrated that the Einstein theory was basically correct.

Other Achievements of Albert Einstein

Folsing [23] discusses in Chapters 29 and 30 the great influence that Einstein had in developing the quantum theory of matter. Einstein received the Nobel Prize for his 1905 analysis of the photoelectric effect. At that time Einstein predicted that light was transmitted in terms of discrete quanta of energy. However, it was not until Arthur Compton's 1923 experiment with X-rays that definite proof was found that electromagnetic waves act like particles. The "Compton effect" proved Einstein's contention that light consists of discrete particles with no rest mass yet with energy and momentum. In 1926 the light quantum was officially recognized and given the name "photon".

In 1924 Einstein cooperated with a young Indian physicist, Saryendra Bose, who had developed a new mathematical method of statistically counting particles. Physicists had counted atoms as if, in principle, they could be numbered and individually identified. Bose's method of statistically counting recognized that atoms are not distinguishable, even in principle.

Einstein supported Bose's basic paper, and immediately applied it in his paper "Quantum Theory of Single-Atom Gasses". Einstein generalized the Bose counting method into a quantum statistics principle that became known as "Bose-Einstein statistics". This new statistical theory enabled Einstein to make remarkable predictions concerning the behavior of matter at extremely low temperatures.

Einstein was impressed by the 1924 finding by Louis De Broglie that particles have wave-like characteristics. Einstein supported De Broglie's PhD thesis and his Nobel Prize in 1929. In September 1924 Einstein recommended that experimenters should search for diffraction and wave interference phenomena in particle beams. By 1927 these wave effects were confirmed by the diffraction of electrons in crystals.

Einstein's Search for a Unified Field Theory

After developing his General theory of Relativity, Einstein devoted most of his efforts in a search for a *Unified Field Theory*. Initially, this would have combined into a single integrated theory the principles of electromagnetic fields and gravitational fields. Later, as knowledge of the atomic nucleus evolved, he also attempted to include the effects of atomic nuclear fields.

Einstein realized that if he could solve this problem he would make a revolutionary advance in physics. He struggled with this task to his last days but never succeeded. A fundamental problem that Einstein faced in this quest is that his General theory of Relativity is not consistent with quantum mechanics. However, the Yilmaz refinement of the Einstein theory is.

Einstein had the wisdom to realize that a *Unified Field Theory* may be achievable, and this would lead to great advances in physics. Since the Yilmaz theory is consistent with quantum mechanics, it may well provide the key for achieving Einstein's elusive *Unified Field Theory*.

APPENDICES

The appendices contain detailed information to support the concepts presented in Chapters 1 to 16. However, one need not read these appendices to understand the material in the chapters. The titles of the appendices are

A. The Marmet redshift effect
B. Analysis of Quasar 3C48
C. Details of Yilmaz cosmology model
    Cosmic microwave background radiation
    Density of matter in the universe
D. The physical meaning of a tensor
E. Relativity analyses
    Formulas for Schwartzschild and Yilmaz solutions
    The meaning of relativistic gravitational potential
    Derivation of Yilmaz theory
    Relativistic Doppler shift
    Einstein pseudo-tensor for the gravitational field
    Conservation of matter-plus-energy
    Einstein's rejection of the Big Bang singularity

Appendix A

The Marmet Redshift Effect

Paul Marmet has derived a redshift effect that he has proposed as an explanation for the Hubble redshift, which is usually interpreted to represent an expansion of the universe. This concept has been strongly supported by Grote Reber, co-initiator of radio astronomy.

Marmet [30, 31] has shown that a photon of light loses energy when it collides with a gas molecule, and thereby experiences a redshift. The direction of the light does not change. This redshift effect cannot be observed in the earth's atmosphere, because the density of gas is too high. It cannot be observed in the laboratory, because it requires too long a path to create a measurable effect.

However, Marmet has shown that his effect can explain the variation of redshift across the disk of the sun for radiation emitted by the sun. The redshift of radiation from the limb of the sun is greater than that from the center. Therefore Marmet has experimental evidence to support his theory.

The Marmet redshift effect is derived in Table A-1. Item (1) gives the formula for the redshift produced by one photon collision with a hydrogen molecule, which is given in Eq. 12 of Ref. [30]. In the discussion following Eq. 19, Marmet gives the estimated blackbody temperature T of the radiation shown in item (2). Applying item (2) to item (1) gives the redshift per collision shown in item (3). Multiplying item (3) by the speed of light (3×10^8 meter/sec) gives the equivalent velocity shift per collision, shown in item (4).

Item (5) gives the general formula for the number of collisions of photons with hydrogen molecules. Parameter D is the density of the gas in molecules per unit volume; L is the path length, and σ is the effective cross section area of a hydrogen atom. Marmet [30] gives in Appendix B, Eq. B8, the value for the cross section (σ) shown in item (6). We consider a path length L of one light year. Item (7) gives the value of one light year in centimeters. For reference we assume in item (8) a gas density of one hydrogen molecule per cubic centimeter. Substituting the values of items (6) to (8) into the formula of item (5) gives the number of photon collisions in item (9). This shows that for a gas density of one hydrogen molecule per cubic centimeter there are 300 photon collisions per light year of path length.

Table A-1: *Calculation of Marmet redshift for a gas density of one hydrogen molecule per cubic centimeter.*

(1) Redshift per collision ($\Delta f/f$) $2.73 \times 10^{-21} \, T^2$
(2) Kelvin temperature T   50,000
(3) Redshift per collision ($\Delta f/f$) 6.83×10^{-12}
(4) Equivalent velocity shift per collision   2.0 mm/sec
(5) Number of collisions $DL\sigma$
(6) Cross section of hydrogen atom (σ) 3.14×10^{-16} cm^2
(7) Path length L (one light year Lyr) 9.47×10^{17} cm
(8) Assumed hydrogen molecule density D   one (H-molecule)/cm^3
(9) Collisions per light year (Lyr)   300 per Lyr
(10) Equivalent velocity shift per light year   600 mm/sec per Lyr
(11) Velocity shift per million light years   600 km/sec per MLyr

The equivalent velocity shift per collision in item (4) is multiplied by the number of collisions per light year in item (9) to obtain the equivalent velocity shift per light year shown in item (10). This is multiplied by one million to obtain the equivalent velocity shift per million light years (MLyr) shown in item (11).

The calculations of Table A-1 are based on a gas density of one hydrogen molecule per cubic centimeter. The velocity shift in item (11) is 30 times our Hubble constant of 20 km/sec per million light years. Therefore a gas density of 1/30 hydrogen atom per cubic centimeter would yield a Marmet redshift effect equivalent to our Hubble constant. This gas density is equal to 33,000 hydrogen molecules per cubic meter.

By examining the motions of galaxies in galaxy clusters, the gravitational effects of all of the matter associated with the clusters has been estimated. As explained in Appendix C, this yields an estimate of 3 hydrogen atoms per cubic meter. The Marmet cosmology theory requires an average gas density of 33,000 hydrogen atoms per cubic meter, which is 10,000 times greater than this value.

Therefore, we conclude that the Marmet redshift effect is probably not adequate to explain the Hubble redshift. On the other hand, as explained in Chapter 11, the Marmet redshift is a strong candidate for explaining the excess redshifts of quasars and certain galaxies.

Appendix B

Analysis of Quasar 3C48

This appendix presents data on the characteristics of quasar 3C48, which was one of the first two quasars discovered, and was analyzed extensively

Analysis by Greenstein and Schmidt

Basic Analysis

The extreme redshift of the quasar was discovered in Feb. 1963 by Maarten Schmidt and Jesse Greenstein. The first quasars that they studied were 3C48 (with a redshift of 0.367) and 3C273 (with a redshift of 0.158). Their analyses of these quasars were reported by their classic 1964 paper in the *Astrophysical Journal.* [35].

Greenstein and Schmidt observed "forbidden" spectral lines of oxygen and neon in the quasar spectra. These forbidden lines are never experienced on earth, and, except for quasars, are observed only in the radiation from gaseous nebula, which are extremely thin clouds of ionized gas. Gas density in these nebulae is usually less than 10^5 electrons per cubic centimeter, whereas the density of a high vacuum on earth is about ten billion times greater than this. To generate these forbidden lines, it is believed that an enormous volume of hot gas of very low density is required. Gaseous nebulae are heated to a temperature of about 10,000 degrees Kelvin by the radiation from stars.

An extensive theory of the processes involved in the generation of forbidden spectral lines has been developed from observations and analyses of gaseous nebula. Greenstein and Schmidt [35] applied this theory to the lines they observed in the spectra of quasars 3C48 and 3C273. They concluded that the forbidden spectral lines of these quasars could not be explained by a viable stellar model that has a large gravitational redshift, and so they stated that quasar redshift must be a Doppler effect produced by an extremely high velocity. With such a high velocity, a quasars must be at an enormous distance, and must radiate an enormous amounts of power.

The theory of forbidden spectral lines evolved from observations of spectra radiated from gaseous nebulae. Most of the observed nebulae have electron densities less than 10^5 electrons per cubic centimeter. The maximum observed electron density was 7×10^6 electrons pet cubic centimeter, which appears to be the upper limit to electron density in a gaseous nebula. The density of the nebula atmosphere can be deduced from the intensity ratios of different spectral lines.

Let us examine the findings that Greenstein and Schmidt [35] derived from their studies of the spectra of quasar 3C48, which has the larger redshift of the two quasars. Based on the presence of forbidden spectral lines and the ratios of the line intensities, Greenstein and Schmidt [35] estimated the electron density N_e of the atmosphere of quasar 3C48 to be 3×10^4 electrons per cubic centimeter. In their Eq. 3b, they estimated that the power radiated in the hydrogen spectral line Hβ per unit volume of gas should be $10^{-25} N_e^2$ erg/sec per cubic centimeter, where N_e is the number of electrons per cubic centimeter. Since 1 watt is 10^7 erg/sec, this can be expressed as

$$P/v = 10^{-32} N_e^2 \text{ watt/cm}^3 \quad \text{(Hβ line)} \quad \text{(B-1)}$$

where P is the power radiated within a volume v of the gas. Setting N_e equal to 3×10^4 electrons gives the following for the power per unit volume radiated by 3C48 in the Hβ spectral line:

$$P/v = 0.9 \times 10^{-23} \text{ watt/cm}^3 \quad \text{(Hβ line, 3C48)} \quad \text{(B-2)}$$

The redshift $\Delta\lambda/\lambda$ of 3C48 is 0.367. Since the velocity ratio V/c is approximately equal to the redshift, Greenstein and Schmidt calculated the velocity V of 3C48 to be 0.367c, which is 110,000 km/sec. They assumed a Hubble constant of 100 km/sec per Mpc (megaparsec), which is 30.7 km/sec per million light years. This gave a distance to 3C48 of 1100 megaparsecs, or 3.6 billion light years.

Based on the measured values of spectral lines of 3C48, Greenstein and Schmidt calculated the power levels that must be radiated in the observed spectral lines of 3C48, assuming that 3C48 is at a distance of

1100 megaparsec, or 3.6 billion light years. These power levels are given in Table B-1. The intensity of spectral line [OIII] was too small to be measured accurately, and so its emitted power is indicated as "present". The lines in brackets [] are forbidden spectral lines, which are not observed on earth. Lines [O II] and [O III] are spectral lines of oxygen, and [Ne V] is a spectral line of neon.

The expression in Eq. B-2 shows that a volume of 7.11×10^{58} cm^3 (or 7.11×10^{43} km^3) is required to generate the power of 6.4×10^{35} watts given for the Hβ line in Table B-1. Since one light year (Lyr) is 9.46×10^{12} km, the gas volume in cubic light years is

$$v = 7.11 \times 10^{43} \text{ km}^3 = 84,000 \text{ Lyr}^3 \tag{B-3}$$

This is equivalent to the volume of a sphere with a radius of 27 light years. Greenstein and Schmidt [35] (p. 1) estimated the gas volume for 3C48 to have a radius of about 10 parsecs (or 32.6 light years), which is reasonably consistent with our calculation.

Table B-1: Power in spectral lines emitted from 3C48, assuming it is at a distance of 3.6 billion light years (1100 megaparsecs)

spectral line	power in watts	power in suns	fraction of total power
Hβ	6.4×10^{35}	1.68×10^9	0.168 %
Mg II	3.1×10^{35}	0.82×10^9	0.082 %
[NeV]	1.7×10^{35}	0.45×10^9	0.045 %
[O II]	3.3×10^{35}	0.87×10^9	0.087 %
[O III]	present		

Greenstein and Schmidt [35] (p. 19) state that the absolute visual magnitude of 3C48 is about -25. Since the absolute visual magnitude of our sun is 4.8 (i.e., about 5), the estimated power radiated from 3C48 is about 30 magnitudes greater than our sun. A magnitude difference of 5 represents a factor of 100 in power. Hence the power radiated from 3C48 should be equivalent to 100^6 suns, or 10^{12} suns. In words this means that the light power radiated from quasar 3C48 should be equal to the power from 1000 billion suns.

For comparison, the power radiated from our Milky Way galaxy is equivalent to 10 billion suns. Therefore the power radiated by the 3C48 galaxy was calculated to be 100 times the power radiated by our

218 *The Scientific Story of Creation*

enormous Milky Way galaxy.

The power radiated by our sun is 3.8×10^{26} watts. In Table B-1, the values of power in watts are divided by the sun power to obtain the power radiated from individual lines in "suns". These values are then divided by the total radiated power from 3C48 (10^{12} suns) to obtain the fraction of the total power contained in the spectral lines, which is shown in the last column.

Rapid Variations of Quasar Brightness

Greenstein and Schmidt [35] (p. 16) report that the optical flux from 3C48 "has changed by a factor of 1.4, apparently independently of wavelength, over a period of 600 days". They do not say whether or not the intensities of the spectral lines varied.

This indicates that 40 percent of the total power, which is the power from 400 billion suns, must be coming from a volume that is no thicker in distance than 600 days. This suggests that the main body of the quasar is only a few light years in diameter, even though it seems to be generating the power of 1000 billion suns.

We can presumably model this quasar as a central core a few light years across, radiating the power of 100 Milky Way galaxies, where this core is surrounded by a thin gas 65 light years in diameter, which is generating the forbidden spectral lines.

On the other hand, if the intensities of the spectral lines vary along with the general power level, this model would be untenable, because the diameter of the thin gas could not be 65 light years in diameter.

Thus we are left with an ambiguous model for this quasar. It is clear that more studies are needed to measure the time variation of power in the forbidden spectral lines.

Other Data on Quasar 3C48

Halton Arp has shown in *Seeing Red* [17] (p. 56, Fig. 2-18) that quasar 3C48 is surrounded by a nebulous fuzz, with a length of about 12 arc seconds. In a personal communication, Halton Arp has reported that the redshift measured from the fuzz differs by no more than 0.001 from the 0.367 redshift value measured for 3C48.

The Marmet redshift effect described in Appendix A allows a more reasonable explanation for the redshift of 3C48. Based on the ratios of the spectral line intensities, Greenstein and Schmidt estimated the atmosphere of 3C48 to have a density of 30,000 electrons per cubic centimeter. Assuming that this is mostly hydrogen, this would represent 30,000 hydrogen molecules per cubic centimeter. For this gas density, the Marmet redshift effect described in Appendix A predicts a redshift equivalent to a velocity of 18 km/sec per light year of path length. This redshift effect would produce the 0.367 redshift of 3C48 with a path length of 6100 light years.

If the Marmet effect is causing the redshift of 3C48 and the associated nebula, the distance to 3C48 can be much less than the 3.6 billion light years calculated by Greenstein and Schmidt. Table B-2 shows the power that would be radiated from the quasar and the length of the associated nebula for different distances to the quasar.

Table B-2: *Power in suns radiated from quasar 3C48 and length of nebulous fuzz for various distances to quasar*

distance (Lyr)	power in suns	nebula length
3.6 billion	1000 billion	210,000 Lyr
360 million	10 billion	21,000 Lyr
36 million	100 million	2,100 Lyr
3.6 million	1 million	210 Lyr

The Marmet redshift effect yields a much more reasonable explanation for the 3C48 quasar. Nevertheless, we are apparently still left with the power of at least one million suns being radiated from a region only a few light years across.

Appendix C

Details of Yilmaz Cosmology Model

Cosmic Microwave Background Radiation

Proponents of the Big Bang theory strongly acclaim the discovery of cosmic microwave background radiation as a milestone in validating the theory. In 1965 this radiation was first detected in a sensitive communication antenna at Bell Laboratories. Much more accurate measurements were obtained from the Cosmic Background Explorer (COBE) satellite in 1989. These experiments show that microwave background radiation is emanating uniformly from all directions, and has the spectrum and intensity that would be emitted from an ideal blackbody at a temperature of 2.73 degrees Kelvin.

Cosmic microwave background radiation was predicted by Big Bang theorists, who claimed it to be the cooled relic of radiation that was emitted from hot plasma 300,000 years after the Big Bang. However estimates of the blackbody temperature predicted by Big Bang theorists varied from 5 °K to 30 °K, and so the cosmic radiation was only predicted in a qualitative sense.

The Yilmaz cosmology model also predicts cosmic microwave radiation. In Chapter 14, Fig. 14-4 shows that at very large distances the apparent galaxy velocity should be very close to the apparent speed of light, and so the light radiated from distant galaxies should be Doppler shifted to very low frequencies. *Believe* [1] presents an analysis of this radiation in Appendix D, which is summarized in this appendix. The predicted radiation is equivalent to the emission from an ideal blackbody at a temperature between 2.1 and 3.4 °K, which agrees closely with the 2.73 °K blackbody temperature measured by the COBE satellite.

The general spectrum of blackbody radiation was shown in Fig. 3-6 of Chapter 3. The wavelength scale was expressed in terms of the half-power wavelength λ_h, which is equal to 4.107/T mm, where T is the

C. Details of Yilmaz Cosmology Model   221

blackbody temperature in degrees Kelvin. Half of the power falls at wavelengths greater than λ_h and half falls at wavelengths less than λ_h. The temperature of a blackbody determines the actual frequencies that it radiates, but the shape of the spectrum is the same for all temperatures.

The light radiated from our sun has a spectrum approximating that of an ideal blackbody at a temperature T of 5770 °K. The corresponding value for the wavelength λ_h is 0.000712 mm, or 0.712 micrometers (millionths of a meter). Our analysis assumes that all stars in our universe have the same spectrum as our sun.

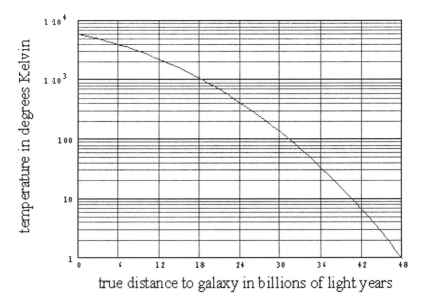

Figure C-1: Blackbody temperature of radiation received from a distant galaxy versus the true distance r to the galaxy

Because the light from a galaxy is Doppler shifted toward lower frequency, the equivalent blackbody temperature of the spectrum that is received from a galaxy decreases with distance to the galaxy. Figure C-1 shows the equivalent blackbody temperature of the Doppler-shifted spectrum that is received from a galaxy at a particular true distance r. This was calculated by applying the Einstein formula for Doppler frequency shift (given in Appendix E) to the data in Fig. 14-4 of Chapter 14, which shows the apparent galaxy velocity divided by the apparent speed of light for the Yilmaz cosmology model. It is the apparent velocity ratio that determines the Doppler frequency shift.

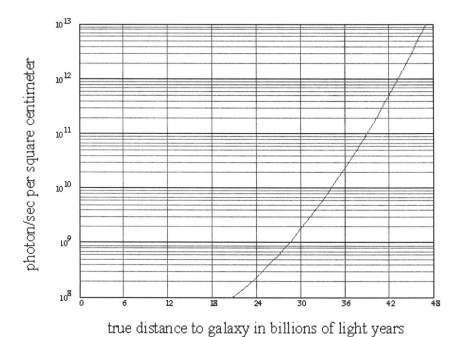

Figure C-2: Photon rate per unit area for the cosmic radiation received from galaxies at a true distance r.

In Fig. C-1 the equivalent blackbody temperature decreases from 5770 °K (the blackbody temperature of the sun) for a close galaxy down to 1 °K for a galaxy at a true distance of 48 billion light years.

Figure 14-3 of Chapter 14 showed that the apparent density of matter should become very high as the apparent distance to a galaxy approaches the limit of 15 billion light years. Because of this very high apparent density, the intensity of the radiation received from a galaxy located at a large true distance should become very high. Figure C-2 shows the photon rate that should be received, per unit area of receiver surface, from galaxies at different values of true distance r. The plot shows that the photon rate becomes extremely large at large true distances.

When we combine the data in Figs. C-1 and C-2, we can plot the intensity of the received radiation versus the equivalent blackbody temperature of the spectrum. This gives the solid curve in Fig. C-3. This shows the photon rate that should fall onto a square centimeter of receiver surface, expressed in terms of the equivalent blackbody temperature T of the received Doppler-shifted spectrum.

C. Details of Yilmaz Cosmology Model 223

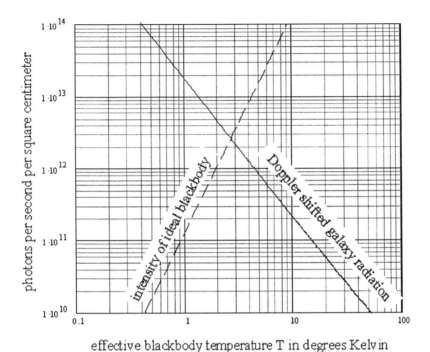

Figure C-3: Photon rate intensity of cosmic radiation received from galaxies, versus effective blackbody temperature of radiation; received radiation (solid); ideal blackbody (dashed)

An ideal blackbody emits a photon rate that is proportional to the cube of the blackbody temperature. The dashed plot in Fig C-3 shows the photon rate per unit area that is emitted from an ideal blackbody versus the temperature of the body.

For an ideal blackbody, the radiation is in thermal equilibrium with the molecules at the surface. We assume that this radiation level cannot be exceeded by cosmic radiation in space. If it were, the diffuse material in space should rapidly absorb the cosmic radiation. Therefore we conclude that the Doppler-shifted galaxy radiation indicated by the solid plot in Fig C-3 cannot exceed the dashed plot. This indicates that the intersection point of the two plots should give the effective blackbody temperature of the received blackbody radiation. The figure shows that this calculated temperature is 2.7 °K. Because of approximations in the analysis, there can be an error in this result. We conservatively estimate that an exact computed temperature should fall within the range from 2.1

°K to 3.4 °K.

Figure C-2 shows that galaxies producing blackbody radiation equivalent to a 2.7 °K temperature are at a distance of 45 billion light years. This indicates that we should receive from galaxies at a true distance of about 45 billion light years cosmic microwave radiation that corresponds to a blackbody at a temperature from 2.1 °K to 3.4 °K.

For comparison, the COBE satellite found that the received cosmic radiation is equivalent to the radiation from an ideal blackbody at a temperature of 2.73 °K. The COBE radiation is received with very high uniformity from all directions, which agrees with our analysis.

Density of Matter in the Universe

The Hubble expansion rate predicted by the Yilmaz cosmology model depends on the average density of matter in the universe. The density is equal to $(3H_0^2/8\pi G)$, where H_0 is the Hubble constant, and G is the gravitational constant of Newton's theory. For a Hubble constant of 20 km/sec per million light years, the average density of matter is equivalent to 4.8 hydrogen atoms per cubic meter.

Big Bang models of the universe define a critical mass density for the universe. According to the Big Bang theory, if the density of matter in the universe is less than the critical density, the universe will expand forever; and if the density of the universe is greater than the critical mass density, the universe will eventually collapse. The critical mass density for the Big Bang theory has the same value $(3H_0^2/8\pi G)$ as the mass density that is required by the Yilmaz theory. This shows that, in order to be valid, the Yilmaz cosmology model requires a density of matter that is equal to the critical mass density for the Big Bang theory.

Believe [1] shows in Chapter 11, Table 11-3, that the estimated density of matter in the universe associated with stars that we can see (called luminous matter) is equivalent to one hydrogen atom per 139 cubic meters. This is a factor of 139x4.8 or 667 times less than critical mass density (4.8 hydrogen atoms per cubic meter). However, there is much more dark matter in the universe (that we cannot see) than there is luminous matter.

For example, there must be about 10 times more dark matter than luminous matter in our Milky Way galaxy for the various parts of the galaxy to rotate with the measured velocities. As shown in *Believe* [1], Appendix F, Table F-1, measurements of gravitational effects indicate that the ratio of dark matter to luminous matter for large galaxy clusters

C. Details of Yilmaz Cosmology Model 225

is typically about 400. This represents an average mass density in the universe of 400/139, or 2.9 hydrogen atoms per cubic meter. This is reasonably close to the density of 4.8 hydrogen atoms per cubic meter that is required by the Yilmaz cosmology model.

What is the source of this dark matter? Big Bang theorists are searching hard for dark matter, because their theories have difficulty explaining the evolution of the universe soon after the Big Bang unless the density of matter is close to the critical density. They have not found sufficient dark matter in the universe to achieve critical mass density.

On the other hand, a fundamental mistake has been made in the Big Bang search for dark matter. Astronomers have used the radiation from quasars to estimate the density of intergalactic gas. Since they assume that quasars are at enormous distances, they have concluded that the density of intergalactic gas must be extremely small. However, quasars are very much closer, and so these estimates of intergalactic gas are not meaningful. Besides, most of the gas in space is probably molecular hydrogen (H_2), rather than atomic hydrogen (H), and molecular hydrogen is very difficult to detect.

It is probable that there is much more hydrogen in the enormous spaces between galaxies than astronomers believe. Hydrogen gas between galaxies may represent the missing dark matter.

If the universe has critical mass density, the total mass in the Big Bang universe is the density ($3H_0^2/8\pi G$) multiplied by the volume of the observable universe, which is $(4/3)\pi r_0^3$. This product is ($H_0^2 r_0^3/2G$). The parameter r_0 is the radius of the observable universe. The expression ($H_0 r_0$) is equal to the speed of light c, and so the total universe mass becomes ($c^2 r_0/2G$). The normalized relativistic mass of the sun, which we denote (m_{sun}), is defined as ($M_{sun}G/c^2$). Hence the total mass of the observable Big Bang universe for critical mass density is equal to

Mass of observable universe = ($c^2 r_0/2G$) = $M_{sun}(r_0/2m_{sun})$

Table 10-1 of Chapter 10 shows in item (3) that m_{sun} is 1.5 km. Chapter 4 shows in Table 4-1, item (5), that r_0 is 142×10^{21} km. Substituting these values into the above equation gives 47×10^{21} M_{sun} for the universe mass.

For critical mass density, the observable Big Bang universe has 47×10^{21} times the mass of our sun. This is also the mass of the Yilmaz cosmology model for a sphere with a radius of 15 billion light years.

Appendix D

The Meaning of a Tensor

Tensor to Specify Stress within a Body

We can illustrate the concept of a tensor by considering the forces inside a mechanical body. A force exerted within a body is specified in terms of stress, which is the force applied per unit area. Two directions are required to describe a stress. One direction gives the direction of the force, and the second direction gives the orientation of the surface to which the force is applied. The stresses exerted within a body are specified in terms of a mathematical concept called a *tensor*.

As an example of the use of tensors, Fig. D-1 shows the forces applied to a cube of material within a body. This body might be solid, liquid, or gas. These forces are expressed in term of x, y, z axes. Each face of this small cube in Fig. D-1 is considered to have unit area. *Force-per-unit-area* is called *stress*, and so the forces applied to the unit-area faces of the small cube are stresses.

These stresses are denoted in the form p_{ab}, where the subscript indices a,b can each represent x, y, or z. The first subscript index (a) describes the direction of the force. The second index (b) describes the orientation of the face to which the force is applied. The orientation of a face is defined by a vector that is perpendicular to the face.

We use the symbol "p" to represent a stress, because pressure is a typical example of a stress. Pressure, which is force per unit area, is specified with English units as pounds per square inch (psi).

Diagram (a) shows the three *compression stresses* that are applied at three faces of the cube, and are denoted p_{xx}, p_{yy}, and p_{zz}. Stress p_{xx} is exerted on the cube in the x-direction at a face that is perpendicular to the x-direction. Stress p_{yy} is exerted in the y-direction at a face that is perpendicular to the y-direction. Stress p_{zz} is exerted in the z-direction at a face that is perpendicular to the z-direction.

D. *Physical Meaning of a Tensor*   **227**

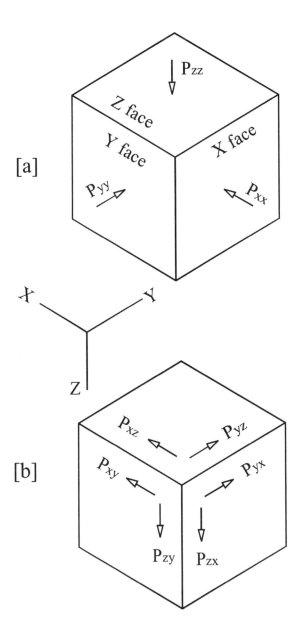

Figure D-1: Internal stresses within a mechanical body:
[a] compression stresses; [b] shear stresses.

The three faces of the cube, which are specified by the second subscript index, are labeled the *x-face*, the *y-face*, and the *z-face*. The x-face is perpendicular to the x-direction, the y-face is perpendicular to the y-direction, and the z-face is perpendicular to the z-direction.

Diagram (b) of Fig D-1 shows the six *shear stresses* that are applied to the faces of the cube. A shear stress is applied parallel to a face. There are two shear stresses applied to the x-face of the cube, which are denoted p_{yx} and p_{zx}. Stress p_{yx} is applied to the x-face in the y-direction, and stress p_{zx} is applied to the x-face in the z-direction. Similarly, the two shear stresses applied to the y-face are denoted p_{xy} and p_{zy}, and the two shear stresses applied to the z-face are denoted p_{xz} and p_{yz}.

Therefore nine separate stresses are required to describe the internal forces within a mechanical body. These stresses are arranged as follows in a 3-by-3 array, which is called a *matrix*

$$\begin{vmatrix} p_{xx} & p_{xy} & p_{xz} \\ p_{yx} & p_{yy} & p_{yz} \\ p_{zx} & p_{zy} & p_{zz} \end{vmatrix}$$

The elements are arranged in this array according to the following rule: the first subscript index indicates the row of an element, and the second index indicates the column. In the first row the first index is x; in the second row the first index is y; and in the third row the first index is z. Similarly the second index is x in the first column; the second index is y in the second column; and the second index is z in the third column.

The compression stresses, shown in diagram (a), are denoted p_{xx}, p_{yy}, and p_{zz}. The compression stresses are located on the diagonal of the matrix, which extends from the upper left to the lower right of the matrix. These are called the "diagonal elements" of the matrix. The other elements (the shear stresses in this case) are called the "nondiagonal elements". In some matrices, the nondiagonal elements are all zero, and the matrix is called a "diagonal matrix".

This matrix is defined by the variable p_{ab}, which is called a *tensor*. This variable p_{ab} represents the nine stresses in a general form. The subscripts a, b are called indices, and each index can represent x, y, or z.

In summary, a tensor with two indices is required to define a stress variable, because stress has two independent directions. The first index specifies the direction of the force, and the second index specifies the orientation of the surface to which the force is applied. A surface is perpendicular to the direction specified by the second index.

D. Physical Meaning of a Tensor

It is often convenient to describe the x, y, z rectangular coordinates by numbers, where x is called x_1, y is called x_2, and z is called x_3. Hence the x index of a stress component is replaced by 1, the y index is replaced by 2, and the z index is replaced by 3. In this form the stress tensor matrix is expressed as

$$\begin{vmatrix} p_{11} & p_{12} & p_{13} \\ p_{21} & p_{22} & p_{23} \\ p_{31} & p_{32} & p_{33} \end{vmatrix}$$

In the stress tensor p_{ab}, the subscript indices a, b now take the values, 1, 2, and 3.

Tensors in Relativity Theory

In Relativity theory, the space and time variables cannot be considered separately. Reality is described by *events*, where an event is specified by three spatial coordinates plus a time coordinate, and so has four dimensions. In Relativity theory a vector has 4 components, and a tensor has 4x4 or 16 components.

In a relativistic tensor, the x, y, z spatial coordinates are specified as x_1, x_2, x_3. Einstein denoted the time coordinate as x_4, but we use the alternative convention where the time coordinate is denoted x_0. Hence we represent a general relativistic tensor by a 4 by 4 array of the form.

$$\begin{vmatrix} p_{00} & p_{01} & p_{02} & p_{03} \\ p_{10} & p_{11} & p_{12} & p_{13} \\ p_{20} & p_{21} & p_{22} & p_{23} \\ p_{30} & p_{31} & p_{32} & p_{33} \end{vmatrix}$$

The first row and the first column give the elements that are associated with the time coordinate.

Appendix E

Relativity Analyses

This appendix presents analyses associated with various aspects of the Einstein and Yilmaz theories.

Formulas for Schwartzschild and Yilmaz Solutions

The following gives the equations for the relativistic effects produced by the gravity of a single star that are predicted by the Schwartzschild solution of the Einstein theory and by the Yilmaz theory.

Table E-1 shows the g_{00} and g_{11} metric tensor elements for the Schwartzschild solution of the Einstein theory and for the single-star solution of the Yilmaz theory. Table E-2 gives general formulas for various effects due to gravity expressed in terms of the g_{00} and g_{11} metric tensor elements. The formulas of these tables are obtained from *Believe* [1], Chapter 8, Tables 8-4 and 8-5.

The g_{00} and g_{11} values from Table E-1 are applied to these general formulas of Table E-2 to obtain the specific formulas for the Einstein Schwartzschild solution and for the Yilmaz solution of a single star. These specific formulas are plotted in Figs. 10-1 to 10-3 of Chapter 10.

Table E-1: *Metric tensor elements for Einstein Schwartzschild solution and Yilmaz solution of the gravitational effects of a star*

	Einstein	Yilmaz
g_{00}	$1 - 2(m/r)$	$\exp[-2m/r]$
g_{11}	$-1/[1 - 2(m/r)]$	$-\exp[2m/r]$

Table E-2: Relativistic effect caused by a gravitational field

Characteristic	General	Einstein	Yilmaz
Speed of light	$\sqrt{[-g_{00}/g_{11}]}$	$[1 - 2(m/r)]$	$\exp[-2m/r]$
Clock rate	$\sqrt{[g_{00}]}$	$\sqrt{[1 - 2(m/r)]}$	$\exp[-m/r]$
Spatial contraction	$1/\sqrt{[-g_{11}]}$	$\sqrt{[1 - 2(m/r)]}$	$\exp[-m/r]$
Wavelength ratio (λ'/λ)	$1/\sqrt{[g_{00}]}$	$1/\sqrt{[1 - 2(m/r)]}$	$\exp[m/r]$

The Meaning of Relativistic Gravitational Potential

The static Yilmaz theory is expressed in terms of the relativistic gravitational potential, which is denoted ϕ. This concept is explained in Table E-3. Item (1) shows that when an object of mass M_o moves at a velocity V, it has a kinetic energy E_k equal to $\frac{1}{2}M_oV^2$. Item (2) shows that when the object is raised through a height h its potential energy E_p is increased by the amount $M_o gh$, where g is the acceleration of gravity. If the object falls through a height h, it loses potential energy, which is converted into kinetic energy. Setting the kinetic energy of item (1) equal to the potential energy of item (2) shows that the velocity V of the object is equal to $\sqrt{[2gh]}$ after it has fallen through the height h.

Table E-3: Calculation of relativistic gravitational potential (ϕ)

(1) Kinetic energy (E_k) of object of mass M_o		$\frac{1}{2} M_o V^2$
(2) Potential energy (E_p) of object		$M_o gh$
(3) Gravitational potential		$(-E_p/M_o) = -gh$
(4) Equivalent energy of mass (E_m)		$M_o c^2$
(5) Relativistic gravitational potential (ϕ)		$(-E_p/E_m) = -gh/c^2$

Gravitational potential is defined as the negative of the change of potential energy E_p of an object divided by its mass. In item (3), the negative of the potential energy E_p of item (2) is divided by the mass M_o of the object to obtain the gravitational potential (-gh).

Gravitational potential is defined in terms of the negative of the potential energy of an object, because gravitational potential describes the gravitational field itself. As height h increases the potential energy of an object increases but the strength of the gravitational field decreases.

Item (4) gives the equivalent energy E_m of an object of mass M_o, which is $M_o c^2$. This is the energy that is released if the mass M_o is converted into energy. As shown in item (5), the relativistic gravitational potential (ϕ) is defined as the negative of the increase of potential energy (E_p) of an object divided by its equivalent energy (E_m).

In item (5), the negative of the potential energy (E_p) of item (2) is divided by the equivalent energy (E_m) of item (4) to obtain the relativistic gravitational potential ϕ. When height is increased by h, the relativistic gravitational potential ϕ changes by $(-gh/c^2)$.

The relativistic gravitational potential for a spherically symmetric body is calculated in Table E-4. Table 5-1 of Chapter 5 showed that the acceleration of gravity at the surface of the earth is GM_e/r_e^2. This can be generalized to show that for any spherically symmetric body of mass M the acceleration of gravity g at any point outside the body is GM/r^2, which is given in item (1) of Table E-4. Item (2) gives the formula for the normalized relativistic mass m of a body of true mass M. Substituting item (2) into item (1) shows that the acceleration of gravity g is equal to mc^2/r^2, as shown in item (1).

Table E-4: *Calculation of the relativistic gravitational potential for a spherically symmetric body*

(1) Acceleration of gravity (g)　　　　　$GM/r^2 = mc^2/r^2$
(2) Normalized relativistic mass (m)　　GM/c^2
(3) Relativistic gravitational potential (ϕ_{12})　$-gh/c^2 = -mh/r^2$
　　　　　　　　　　　　　　　　　　　　$= -m(\Delta r/r^2)$
(4) From calculus, $(\Delta r/r^2)$ is　　　　$-\Delta(1/r)$
(5) Relativistic gravitational potential (ϕ_{12})　$m[\Delta(1/r)] = \Delta(m/r)$
　　　　　　　　　　　　　　　　　　　　$= [(m/r_2) - (m/r_1)]$
(6) Approximate (V_{12}/c)　　　　　　　$gh/c^2 = -\phi_{12}$

In the elevator of Fig. 8-1b in Chapter 8, denote the change of relativistic gravitational potential from point (1) to point (2) as ϕ_{12}. Item (5) of Table E-3 shows that ϕ_{12} is equal to $-gh/c^2$, as given in item (3) of Table E-4. Substituting into item (3) the expression for g in item (1), shows that ϕ_{12} is equal to $(-mh/r^2)$. Height h represents an increase of radius r, and so can be expressed as Δr. Hence item (3) shows that ϕ_{12} is equal to $-m(\Delta r/r^2)$.

As shown in item (4), one can prove with calculus that $(\Delta r/r^2)$ is equal to $-\Delta(1/r)$, which represents the negative of the change in $(1/r)$.

Substituting item (4) into item (3) shows that ϕ_{12} is equal to $m[\Delta(1/r)]$, as shown in item (5). This can be expressed as $\Delta(m/r)$, which means the change of the ratio m/r. At point (1) m/r is (m/r_1), and at point (2) m/r is (m/r_2). As shown in item (5), the change of m/r from point (1) to point (2) is $[(m/r_2) - (m/r_1)]$. This is the relativistic gravitational potential change ϕ_{12} from point (1) to (2) in the Fig. 8-1b elevator. The elevator could be located on earth or on any spherical symmetric body.

Table 8-1 of Chapter 8 shows that there is an effective velocity difference ΔV between points (1) and (2) approximately equal to (gh/c). We denote ΔV as V_{12}. The velocity ratio V_{12}/c is approximately equal to (gh/c^2), as shown in item (6) of Table E-4. Comparing items (3) and (6) shows that V_{12}/c is approximately equal to $(-\phi_{12})$.

Derivation of Yilmaz Theory

Let us use the relations of Table E-4 to calculate an exact formula for gravitational redshift, in accordance with the Yilmaz analysis. In the elevator of Fig. 8-1 in Chapter 8, assume that a point (p) lies between points (1) and (2). The velocity change from point (1) to point (p) is denoted V_{1p}, the velocity change from point (p) to point (2) is denoted V_{p2}, and the velocity change from point (1) to point (2) is denoted V_{12}. The following well-known formula of Special Relativity shows how these three velocities are related.

$$(V_{12}/c) = \frac{(V_{1p}/c) + (V_{p2}/c)}{1 + (V_{1p}/c)(V_{p2}/c)} \qquad (E-1)$$

This must hold for any possible position of point (p) between points (1) and (2). By item (6) of Table E-4, if points (1) and (2) are very close together, (V_{12}/c) must approximate $(-\phi_{12})$. This means that $(-\phi_{12})$ is the limit of V_{12}/c, as point (2) approaches point (1). Equivalent limits must also hold for V_{1p}/c and V_{p2}/c. The only general answer having the same form for all three velocities that satisfies Eq. E-1 and these limits is

$$(V_{12}/c) = \tanh[-\phi_{12}] \qquad (E-2)$$

$$(V_{1p}/c) = \tanh[-\phi_{1p}] \qquad (E-3)$$

$$(V_{p2}/c) = \tanh[-\phi_{p2}] \qquad (E-4)$$

234 The Scientific Story of Creation

The expression $\tanh[-\phi_{12}]$ is called the hyperbolic tangent of $(-\phi_{12})$. The hyperbolic tangent is defined by

$$\tanh[x] = (e^x - e^{-x})/(e^x + e^{-x}) \tag{E-5}$$

For very small values of x, $\tanh[x]$ is approximately equal to x.

The exact Doppler shift formula was derived by Einstein in Special Relativity. It is given below in Eq. E-21 and can be expressed as

$$(\lambda_2/\lambda_1)^2 = \frac{1 + (V_{12}/c)}{1 - (V_{12}/c)} \tag{E-6}$$

Let us represent $(-\phi_{12})$ as x, so that (V_{12}/c) in Eq. E-2 is equal to $\tanh[x]$. Apply the expression for $\tanh[x]$ in Eq. E-5 to Eq. E-6. This gives

$$(\lambda_2/\lambda_1)^2 = \frac{1 + \tanh[x]}{1 - \tanh[x]} = \frac{1 + (e^x - e^{-x})/(e^x + e^{-x})}{1 - (e^x - e^{-x})/(e^x + e^{-x})}$$

$$= \frac{(e^x + e^{-x}) + (e^x - e^{-x})}{(e^x + e^{-x}) - (e^x - e^{-x})} = e^{2x} \tag{E-7}$$

Taking the square root gives

$$\lambda_2/\lambda_1 = e^x = \exp[x] = \exp[-\phi_{12}] \tag{E-8}$$

Applying the expression for ϕ_{12} in item (5) of Table E-4 gives

$$\lambda_2/\lambda_1 = \exp[-\phi_{12}] = \exp[(m/r_1) - (m/r_2)] \tag{E-9}$$

This is an exact formula that holds for all values of the radial distances r_1 and r_2. Let us consider this wavelength ratio when r_2 is infinite. The wavelength λ_2 is then called λ'. Making r_2 infinite in Eq. E-9 gives

$$\lambda'/\lambda_1 = \exp[m/r_1] \tag{E-10}$$

We can replace λ_1 and r_1 by λ and r, and the equation simplifies to

E. Relativity Analyses   235

$$\lambda'/\lambda = \exp[m/r] = e^{m/r} \tag{E-11}$$

Table E-2 shows that the wavelength ratio (λ'/λ) is equal to $1/\sqrt{[g_{00}]}$. Hence, by Eq. E-11, the metric tensor element g_{00} is equal to

$$g_{00} = 1/(\lambda'/\lambda)^2 = 1/(e^{m/r})^2 = e^{-2m/r} \tag{E-12}$$

To obtain the other metric tensor elements, Yilmaz postulated that the speed of light measured locally is independent of direction. Yilmaz has proven that to satisfy this requirement the following conditions must hold in rectangular coordinates:

$$g_{00}g_{11} = -1 \ ; \quad g_{33} = g_{22} = g_{11} \tag{E-13}$$

The first equation is derived in *Addendum* [3], Chapter 3, Section 3.5. The second equation should be obvious. Applying Eq. E-13 to Eq. E-12 gives the following elements for the metric tensor:

$$g_{00} = e^{-2m/r} \tag{E-14}$$

$$g_{11} = g_{22} = g_{33} = -e^{2m/r} \tag{E-15}$$

This gives the metric tensor elements for the static Yilmaz theory.

These metric tensor elements apply to the gravitational field of a single body having a spherical symmetric mass distribution. The formulas can be generalized by expressing them in terms of the relativistic gravitational potential ϕ. For a spherically symmetric body having normalized relativistic mass m, the relativistic gravitational potential ϕ is equal to m/r at any point outside the body. Replacing (m/r) in the above equations by the relativistic gravitational potential (ϕ), gives the following general formulas for the static Yilmaz theory:

$$g_{00} = e^{-2\phi} \tag{E-16}$$

$$g_{11} = g_{22} = g_{33} = -e^{2\phi} \tag{E-17}$$

These metric tensor elements can characterize the gravitational field for any number of bodies. At any point external to several spherically symmetric bodies, the gravitational potential is

$$\phi = (m_1/r_1) + (m_2/r_2) + (m_3/r_3) + \text{etc.} \tag{E-18}$$

Variable r_1 is the distance measured from the center of body (1) having mass m_1. The other quantities are defined accordingly. From Eqs. E-16 to E-18 one can readily calculate the relativistic gravitational interactions of all of the bodies of our solar system.

Relativistic Doppler Shift

In 1842, Christian Doppler (1803-1853), an Austrian scientist, showed that the frequency of a sound or light wave is changed by the relative velocity between the emitter and receiver. Einstein proved that the Doppler formulas are different for sound and light, because sound travels relative to the air, whereas relativistic effects are involved in light transmission. In his 1905 paper on relativity, Einstein [9] derived the following relativistic formula for the Doppler shift of a light wave:

$$f'/f = \lambda/\lambda' = \frac{1 - (V/c)}{\sqrt{[1 - (V/c)^2]}} \tag{E-19}$$

Variable f is the emitted frequency and f' is the received frequency. The frequency ratio f'/f is equal to λ/λ', where λ is the emitted wavelength and λ' is the received wavelength.

Equation E-19 is given in Ref. [9] (Section 7, page 56, fifth equation on the page). The relative velocity V is assumed to be radial, so that the angle ϕ in the Einstein equation is set to zero. Frequencies f, f' are denoted ν, ν' in the Einstein equations.

It can be shown that Eq. E-19 can be expressed in the following alternate form

$$f'/f = \lambda/\lambda' = \frac{\sqrt{[1 - (V/c)^2]}}{1 + (V/c)} \tag{E-20}$$

Squaring Eq. E-20 and taking the reciprocal yields

$$(\lambda'/\lambda)^2 = (f/f')^2 = \frac{1 + (V/c)}{1 - (V/c)} \tag{E-21}$$

One can prove with algebra that Eqs. E-19 to E-21 are equivalent by noting that $(1 - x^2)$ is equal to $(1 - x)(1 + x)$. Solving Eq. E-21 for V/c gives

$$\frac{V}{c} = \frac{(\lambda'/\lambda)^2 - 1}{(\lambda'/\lambda)^2 + 1} \tag{E-22}$$

Astronomers use this formula to calculate the relative velocity V of a star or galaxy from the wavelength ratio λ'/λ. The received wavelength λ' is defined as $(\lambda + \Delta\lambda)$, where $\Delta\lambda$ is the shift in wavelength produced by the Doppler effect. For small wavelength shifts, Eq. E-22 can be approximated by

$$V/c \cong \Delta\lambda/\lambda \tag{E-23}$$

The symbol $\cong$ means *approximately equal to*. For a receding velocity, $\Delta\lambda$ and V are positive, and the ratio $\Delta\lambda/\lambda$ is called the *redshift*.

Einstein Pseudo Tensor for the Gravitational Field

Einstein recognized the desirability of a stress-energy tensor to describe the gravitational field. He searched for a tensor to characterize the energy of the gravitational field, but could not isolate a true tensor for this purpose. Einstein found what is called a "pseudo-tensor" to describe the energy of the gravitational field. However, he could not use this in his gravitational field equation because it was not a true tensor.

As explained by Pauli [4] (p. 176), Einstein believed that his pseudo-tensor represented the "energy components of the gravitational field". Einstein denoted his pseudo-tensor as t_a^b. To distinguish this Einstein pseudo tensor from the corresponding true tensor of the Yilmaz theory, we denote the Einstein pseudo-tensor in bold format with a strike-through as $\bf{\not{t}}_a^b$.

Yilmaz has proven that the Einstein pseudo-tensor is related as follows to the corresponding true tensor of the Yilmaz theory

$$4\pi \bf{\not{t}}_a^b = -t_a^b + z_a^b \tag{E-24}$$

The variable $\bf{\not{t}}_a^b$ is the Einstein pseudo-tensor for the gravitational field, t_a^b is the Yilmaz stress-energy tensor for the gravitational field, and z_a^b is a non-tensor, which varies with the coordinate system. Although t_a^b is a true tensor, the non-tensor component z_a^b corrupts the Einstein pseudo-tensor $\mathbf{t}_a^b$, thereby keeping the Einstein pseudo-tensor from behaving like a true tensor.

A crucial requirement that Einstein faced in developing his gravitational field equation is that all terms of his equation must be true tensors. As was shown in Chapter 7 in our discussion of Special Relativity, the "*Principle of Covariance*" must be satisfied in a Relativity theory, so that the same laws of physics apply in all coordinate systems. To satisfy this principle when Relativity was generalized, Einstein constrained his gravitational field equation to consist only of true tensors. Consequently Einstein could not use his pseudo tensor in his gravitational field equation.

Tensor theory provides a precise formula for transforming tensors into different coordinates. If physical laws are expressed in terms of true tensors, the laws hold in all coordinate systems.

Conservation of Matter-Plus-Energy

In order to yield realistic predictions, a relativistic theory must achieve conservation of matter-plus-energy. Since matter and energy can be converted into one another, it is the sum of the two that must be conserved.

The *Bianchi identity* is an important theorem of tensor analysis. This identity states that the *covariant derivative* of the Einstein tensor G_a^b is always zero. (The *covariant derivative* is explained in *Believe* [1], Appendix J.) Since the Einstein gravitational field equation sets the energy-momentum tensor T_a^b proportional to the Einstein tensor G_a^b, the *covariant derivative* of the energy-momentum tensor T_a^b is always zero in the Einstein theory.

It is commonly believed that conservation of matter-plus-energy is assured if the *covariant derivative* of the energy-momentum tensor T_a^b is zero. However, the Russian authors Landau and Lifshitz state forcefully in their highly respected book on Relativity theory [65] (page 180) that this common belief is not true. They demonstrate that the *derivative of the tensor-density* of the energy-momentum tensor must be zero in order to achieve conservation of matter-plus-energy.

The reader need not be concerned with the mathematical definitions of the *covariant derivative* and the *derivative of the tensor-density*. It is sufficient to recognize that these constraints are mathematically different and therefore are inconsistent. If the *covariant derivative* of the energy-momentum tensor is zero, the *derivative of its tensor-density* generally cannot be zero. Therefore, the Einstein theory generally cannot achieve conservation of matter-plus-energy, despite the common belief that it does.

As shown in *Addendum* [3], Chapter 5, the *"Freud identity"* assures that the *derivative of the tensor-density* of the energy-momentum tensor is always zero in the Yilmaz theory. Consequently the Yilmaz theory always achieves conservation of matter-plus-energy.

This issue is analyzed in mathematical terms in *Addendum* [3], Chapter 3, Section 3.6.

This discussion has shown that the energy-momentum tensor for the Yilmaz theory always achieves conservation of matter-plus-energy, but that of the Einstein theory generally does not. Therefore the energy-momentum tensors for the two gravitational theories are usually different, even though calculated from the same physical model.

Einstein's Rejection of the "Big Bang" Singularity

In 1945, Einstein [66] (p. 129) rejected the concept that our universe began as a "singularity" with the following statement:

"The theoretical doubts [concerning the start of the universe expansion] are based on the fact that for the time of the beginning of the expansion, the metric becomes singular and the density becomes infinite. In this connection, the following should be noted: The present theory of relativity is based on a division of physical reality into a metric field (gravitation) on the one hand, and into an electromagnetic field and matter on the other hand. In reality, space will probably be of a uniform character, and the present theory will be valid only as a limiting case. For large densities of field and of matter, the field equations and even the field variables which enter into them will have no real significance. One may not therefore assume the validity of the equations for very high density of field and of matter, and one may not conclude that the 'beginning of the expansion' must mean a singularity in the mathematical sense. All we have to realize is that the equations may not be continued over such regions."

GLOSSARY

See Reference [26] for more definitions relating to astronomy.

numbers, exponential representation
  10^5 means 1 followed by 5 zeros; 10^{-5} means $1/10^5$.
  For example, $3.1 \times 10^5 = 3.1 \times 100,000 = 310,000$.
  $3.1 \times 10^{-5} = 3.1/100,000 = 0.000,031$
number prefixes:
  nano = billionth $(1/10^9)$ mega = million (10^6)
  micro = millionth $(1/10^6)$ kilo = thousand (1000)
  milli = thousandth (1/1000)
one billion = 1000 million; one trillion = 1000 billion

acceleration: The rate-of-change of velocity with time.

acceleration of gravity g: Rate of increase of velocity when a body is allowed to fall to earth; equal to 9.8 meter/second per second at earth surface.

aether (also called ether): Postulated medium that provides an absolute reference relative to which light propagates; Einstein rejected the *aether* concept.

apparent age of the universe: The time elapsed since the Big Bang if the universe always expanded uniformly at the present Hubble constant; equal to 15 billion years for a Hubble constant of 20 km/sec per million light years.

archaea: Simple cells without nuclei, but distinct from bacteria; see pps. 14-15.

Big Bang theory: The hypothesis that our universe began as a highly dense mass that exploded with a Big Bang about 15 billion years ago.

black dwarf star: The cold, dark remains of a white dwarf star after it stops radiating energy; the ultimate state of our sun.

black hole: The Einstein theory predicts that a star with a mass-to-radius ratio exceeding 240,000 times that of our sun must collapse to form a *black-hole singularity* having zero size and infinite mass density; light cannot escape from within the *event horizon* sphere that surrounds a black hole.

blueshift: Wavelength shift to shorter wavelength, defined as wavelength decrease divided by normal wavelength; proportional to star velocity toward the earth.

brown dwarf star: A dark star that has less that 1/12 of the sun mass and so is too small to ignite nuclear fusion.

Cepheid variable star: A star that varies periodically in power; since radiated power varies with frequency, this star gives a measure of its distance.

coordinates: For 3-dimensional spatial measurements: *rectangular (or Cartesian) coordinates* use 3 distances; *spherical coordinates* use distance plus 2 angles. *Relativity coordinates* have a fourth dimension that represents time.

cosmology: The study of the evolution and large-scale structure of our universe.

covariance, principle of: Expressing a law of physics such that it is independent of the velocity, acceleration and gravitational field at the location of the observer.

Glossary 241

diagonal elements of a tensor: Those 4 elements for which the two indices are equal. The other 12 elements are called *nondiagonal*.

Doppler wavelength shift: Wavelength change due to velocity; wavelength change divided by normal wavelength is approximately equal to radial velocity divided by the speed of light. Appendix E gives exact formula.

Einstein gravitational field equation: Tensor formula specifying the Einstein General theory of Relativity; which represents 10 independent equations.

electromagnetic field equations: Mathematical formulas derived by Maxwell, which specify the physical laws of electricity and magnetism.

electromagnetic wave: Light wave, radio wave, etc., which consists of electric and magnetic fields oscillating at right angles to one another.

equivalent energy of matter: Energy equal to Mc^2 that would be released if a mass M were converted into energy.

ether: See *aether*.

eukaryote: Cell with nucleus; all multi-celled organisms consist of eukaryote cells.

event horizon: Spherical surface surrounding a *black-hole* over which speed of light is zero; Einstein theory predicts light cannot escape through this surface.

galaxy: A group of many billons of stars; nearly all of the stars of the universe are parts of galaxies; but some are parts of smaller groups called *stellar clusters*.

galaxy, types of: Milky Way galaxy, with the shape of a disk, is a *spiral galaxy*. The other major type is the *elliptical galaxy*, having only an elliptical nucleus.

gravitational constant G: The constant of proportionality in Newton's law of gravitational attraction; defined in Chapter 5.

gravitational potential: The negative of the change of gravitational energy of an object divided by its mass M; which is equal to (-gh), where (g) is the acceleration of gravity and (h) is the change in height. (See App. E.)

gravitational potential, normalized relativistic: Negative of gravitational energy of object divided by its equivalent energy (Mc^2); equal to $(-gh/c^2)$. (See App. E.)

Hubble Constant: The ratio of receding galaxy velocity to galaxy distance for distant galaxies; the latest value is about 20 km/sec per million light years.

Hubble expansion: The 1929 finding by Hubble that galaxies are receding at velocities approximately proportional to distance.

Hubble Law: The postulate that our whole universe expands at a constant rate.

Kinetic energy: The energy of an object of mass M due to its velocity V; ½ MV^2.

Light year: Distance light travels in one year; 9.46×10^{12} kilometers.

Lorentz transformation: The equations specifying Special Relativity, originally derived by Lorentz, which describe the changes of distance and time measurements of two observers moving at different velocities.

magnitude: Logarithmic measure of light power received from a star; the dimmer the star, the greater the magnitude; power decreases by 2.5 for a magnitude change of +1, and by 100 for a magnitude change of +5.

magnitude, absolute: The magnitude a star would have if observed at a distance of 10 parsecs or 32.6 light years; the absolute magnitude of our sun is 4.8.

Marmet redshift: The prediction by Paul Marmet that a hydrogen cloud produces a redshift proportional to gas density and cloud thickness. (See Appendix A)

mass, rest: Mass of object varies with velocity; *rest mass* is mass at zero velocity.
momentum: The product of mass times velocity.
momentum, angular: For mass elements rotating about a center of gravity, the sum (for each element) of momentum times radial distance from the center.
nebula: Originally a fixed nebulous object, including galaxies and stellar clusters; now restricted to gas clouds heated by star radiation, called *gaseous nebulae.*
neutron star: A star having the maximum mass density allowable by physical laws, consisting of tightly packed neutrons; density is 300 million tons per cubic cm.
nuclear fission: Energy released by splitting atom in atomic bomb and nuclear power plant.
nuclear fusion: Energy released by combining atoms; in the sun and the hydrogen bomb.
parallax: the relative image shift of a nearby object due to motion of the observer.
parsec: Theoretical distance of a star exhibiting an annual parallax shift of ±1 arc second because of rotation of the earth around the sun; equal to 3.26 light years.
potential energy: Increase of energy (Mgh) of mass M if raised through height h.
proper coordinates: The *proper coordinates* of an object are located at the object and move with the object; *proper distance* and *proper time* are measurements made on the object relative to these coordinates.
pulsar: A star that emits radio pulses at precise intervals, typically at 10 to 1000 pulses per second; a rapidly spinning neutron star.
quasar or quasi-stellar object (QSO): A star-like object with a very large redshift.
red giant star: State our sun will reach in 5 billion years after converting nearly all of its hydrogen to helium; sun will swell to half the distance to the earth.
receding velocity: Velocity component away from earth.
redshift: Wavelength increase of spectral lines divided by normal wavelength.
redshift, Doppler: Redshift due to velocity; approximately equal to receding velocity divided by speed of light. (Appendix E gives exact formula.)
redshift, gravitational: Redshift produced by a gravitational field.
redshift, intrinsic: Component of galaxy or quasar redshift unrelated to velocity.
Relativity, Special theory of: Einstein's basic theory (1905), which shows how time and spatial measurements are changed by the velocity of the observer.
Relativity, General theory of: Generalization (1916) of Einstein's Relativity theory to include effects of acceleration and gravity.
Schwartzschild limit: The maximum mass-to-radius ratio of a star for which the Schwartzschild solution has a real answer; 240,000 times the ratio for our sun.
Schwartzschild solution: The first exact analytical solution (1916) derived from Einstein General Relativity, specifying gravitational effects of an ideal star having constant density and no viscosity; used to verify Einstein theory.
singularity: Big Bang prediction where size shrinks nearly to zero without change of mass, so that density of matter is nearly infinite.
solar system: the planets, comets, and other bodies orbiting a star.
spectral lines: Most of the light from a star has a continuous *spectrum* like the blackbody in Fig. 3-6. A portion consists of discrete *spectral lines* produced by elements. Some are bright emission lines, and some are dark absorption lines.

spectrum: Pattern formed by passing light through a prism to separate wavelengths.
speed of light: 300,000 kilometers per second (denoted c).
stars, types of (See individual types)
  black dwarf, brown dwarf, neutron star, pulsar, red giant, white dwarf.
Steady-State Universe theory: In 1948 Hoyle, et al, postulated that universe age is infinite, and diffuse matter is created to compensate for Hubble expansion.
stellar: pertaining to a star.
supernova, An enormous stellar explosion equivalent to hundreds of millions of suns; the end of life of a massive star, or effect of white dwarf sucked into a star.
sun parameters: radius = 696,000 km; mass = 1.989×10^{30} kg; density = 1.4; normalized mass m = 1.475 km
temperature, scales of: Celsius (or Centigrade) scale is zero °C at freezing point of water and 100 °C at boiling point; Kelvin scale has Celsius intervals but is zero at absolute zero (-273.15 °C), where random molecular motion is zero.
tensor definition: A generalized variable, usually with 16 elements in Relativity theory; but the high-order Riemann tensor has 256 elements.
tensor, types of:
  Einstein tensor, Denoted $G_a{}^b$ in usual form, describes the curvature of space; closely related to the Ricci tensor.
  energy-momentum tensor: Denoted $T_a{}^b$ or T^{ab} in usual forms, describes the properties of matter and energy.
  metric tensor: Denoted g_{ab} or g^{ab} in usual forms, describes the shortest distance between two points in curved space.
  Ricci tensor: Denoted R_{ab} or $R_a{}^b$ in usual forms, describes curvature of space; closely related to Einstein tensor, and derived from Riemann tensor.
  Riemann tensor, Uniquely describes curvature of space; has 4 indices and 256 elements; not used directly in Relativity calculations.
  stress-energy tensor for gravitational field: The Yilmaz tensor $t_a{}^b$ specifies energy and stress of gravitational field (not used in Einstein theory).
  stress-energy tensor for matter: The name Yilmaz gives for his energy-momentum tensor; denoted $\tau_a{}^b$ and equal to $4\pi T_a{}^b$.
tensor, forms of: A tensor can have three separate forms; the *covariant* form has subscript indices; the *contravariant* form has superscript indices, and the *mixed* form has one subscript and one superscript index.
tensor, diagonal: This class of tensor has only 4 nonzero elements, which are on the diagonal of the tensor matrix (both indices are equal).
vector: Variable having amplitude and direction, represented by an arrow.
velocity, radial or tangential: Radial velocity is velocity in the radial direction, toward or away from earth; *tangential velocity* is perpendicular to the radius.
white dwarf: When nuclear fuel is depleted, our sun will shrink to a *white dwarf*, glowing white hot from energy released by gravity; after reaching the size of the earth, it will cool to become a *black dwarf*.
Yilmaz theory of gravity: Refinement of the Einstein General theory of Relativity, which has achieved an exact solution to the principles of the Einstein theory.

BIBLIOGRAPHY

There are two sets of references. The preface Y indicates references on the Yilmaz theory.

Yilmaz Theory Bibliography

[Y1] Huseyin Yilmaz, "New Approach to General Relativity", *Physical Review*, vol. 111, No. 5, Sept. 1, 1958, pp 1417-1426,
[Y2] Huseyin Yilmaz, "New Theory of Gravitation", *Physical Review Letters*, vol. 27, No. 20, 15 Nov. 1971, pps. 1399-1402.
[Y3] Huseyin Yilmaz, "New Approach to Relativity and Gravitation", *Annals of Physics*, Academic Press, NY, 1973, pps. 179-200.
[Y4] Huseyin Yilmaz, "New Theory of Gravitation", *Proc. 4th Marcel Grossman Meeting Gen. Relativity*, Remo Ruffini, ed, Rome Univ, Italy, June 1985.
[Y5] Huseyin Yilmaz, "New Direction in Gravity Theory", *Hadronic Journal*, 1986, vol. 9 No 6, pp 281-291,.
[Y6] Huseyin Yilmaz, "Present Status of Gravity Theories", *Hadronic Journal*, 1986, vol. 9 No 6, pp 233-238.
[Y7] Huseyin Yilmaz, "Dynamics of Curved Space", *Hadronic Journal*, 1986, vol. 9 No 2, pp 55-60.
[Y8] H. Yilmaz, "Toward a Field Theory of Gravitation", *Nuovo Cimento, B Gen. Physics*, 1992, vol. 107, Iss. 8, pp 941-960.
[Y9] Yilmaz, Huseyin, "Did the Apple Fall?", in *Frontiers of Fundamental Physics*, 1994, M. Barone and F. Selleri, eds, pp. 115-124, Plenum Press, NY.
[Y10] Alley, Carroll O., "Investigation with lasers, atomic clocks [etc.] of gravitational theories of Yilmaz and Einstein", in *Frontiers of Fundamental Physics*, 1994, M. Barone and F. Selleri, eds, pp. 125-137, Plenum Press, NY.
[Y11] Huseyin Yilmaz, "Gravity and Quantum Field Theory, a Modern Synthesis", *Ann New York Acad Science*, 1995, vol 755, pp 476-499.
[Y12] Carroll O. Alley, "The Yilmaz Theory of Gravity and its Compatibility with Quantum Theory", *Ann New York Acad Science*, 1995 vol 755, pp 464-477.

General Bibliography

[1] Adrian Bjornson, *A Universe that We Can Believe*, Addison Press, Woburn, MA, 2000, ISBN 09703231-0-7.
[2] Adrian Bjornson, *How Was Our Universe Creastd?*, Addison Press, Woburn, MA, 2000, ISBN 09703231-1-5.
[3] Adrian Bjornson, *Addendum to, "A Universe that We Can Believe"*, available at no cost on website www.olduniverse.com.

Textbooks on Relativity
[4] W. Pauli, *Theory of Relativity*, 1958 Pergammon Press, reprint Dover Pub, NY, ISBN 0-486-64152-X
[5] Tullio Levi-Civita, *The Absolute Differential Calculus*, 1977, Dover Pub, NY, (first Italian ed, 1923), ISBN 0-486-63401-9.
[6] John A. Peacock, Cosmological Physics, 1999, Cambridge U. Press, United Kingdom, ISBN 0-521-42270-1.
[7] Richard A. Mould, *Basic Relativity*, Springer, NY, 1994, ISBN 0-387-94188-6.

Classic Papers on Relativity
[8] H. A. Lorentz, "Electromagnetic Phenomena in a System Moving with any Velocity Less than that of Light", *Proc. Acad Science Amsterdam*, 1904, vol 6, reprint in *The Principle of Relativity*, Dover Pub, NY, 1952, pp. 9-34.
[9] Albert Einstein, "On the Electrodynamics of Moving Bodies", Annalen der Physik, 1905, vol. 17, English transl. in *The Principle of Relativity*, 1952, Dover Pub, NY, pp. 35-71.
[10] Albert Einstein, "The Foundation of the General Theory of Relativity", Annalen der Physik, vol. 49, 1916, English transl. in *The Principle of Relativity*, 1952, Dover Pub, NY, pp. 109-164.
[11] G. Ricci and T. Levi-Civita, "Methods de calcul differential absolu et leurs applications", *Math. Ann.*, 1901, vol. 54, pp. 125-201.
[12] J. R. Oppenheimer and H. Snyder, "On Continued Gravitational Contraction", *Physical Review*, Sept. 1939, vol 56, pp 455-459.
[13] Albert Einstein, "On a stationary system with spherical symmetry consisting of many gravitating masses", *Annals of Mathematics*, Oct. 1939, vol 40, No 4, pp 922-936 (see p. 936).

Books on Cosmology
[14] George Gamow, *One, Two, Three . . . Infinity*, Bantam Books, 1967, original Viking edition, 1947.
[15] George Gamow, *A Star Called The Sun*, Bantam Books, 1970, original Viking edition, 1964.
[16] Halton C. Arp, *Quasars, Redshifts, and Controversies*, 1987, Interstellar Media, Berkeley, Calif, ISBN 0-941325-00-8.
[17] Halton C. Arp, *Seeing Red*, 11998, Aperion, Montreal, Quebec, ISBN 0-9683689-0-5. (available at Internet website *www.Amazon.com*)
[18] Eric Lerner, *The Big Bang Never Happened*, Times Books div Random House, NY, 1991, ISBN 0-8129-1853-3.
[19] David Filkin, *Stephen Hawking's Universe, the Cosmos Explained*, 1997, Basic Books div Harper Collins, NY, ISBN 0-465-08199-1.
[20] Fred Hoyle, Geoffrey Burbidge, and Jayant Narlikar, *A Different Approach to Cosmology*, 2000, Cambridge U. Press, England, ISBN *0-521-66223-0*.
[21] Joseph Silk, *The Big Bang*, 1989, W. H. Freeman, NY, ISBN 0-7167-1812-X.

[22] Joseph Silk, *A Short History of the Universe*, 1994, Scientific American Library, W. H. Freeman, NY, ISBN-0-7167-5048-1.

Book on Albert Einstein
[23] Albrecht Folsing, *Albert Einstein, a Biography,* 1997, (transl. from German by Ewald Osers). Penguin Books, NY, ISBN 0-14-02.3719-4.

Books on Astronomy
[24] Donald Goldsmith, *The Astronomers*, 1991, St. Martin Press, NY, ISBN 0-312-05380-0.

[25] Kevin Krisciunas and Bill Yenne, *The Pictorial Atlas of the Universe*, 1989, Mallard Press, ISBN 0-792-45200-3.

[26] Valerie Illingworth & John Clark, *Dictionary of Astronomy*, 4th ed, Checkmark Books, NY, ISBN 0-8160-4284-5.

Papers on Astronomy and Cosmology
[27] Geoffrey Burbidge, "Why Only One Big Bang", *Scientific American*, February, 1992, page 120.

[28] Peebles, Schramm, Turner, and Kron, "The Evolution of the Universe", *Scientific American*, Oct. 1994, pps 53-65.

[29] Ann Finkbeiner, Astronomy: Hubble Telescope Settles Cosmic Distance Debate, or Does it?", *Science*, May 28, 1999.

[30] Paul Marmet, "A New Non-Doppler Redshift", presented in Internet website: www.newtonphysics.on.ca.

[31] Paul Marmet, "Discovery of H_2 in space explains dark matter and redshift", originally published in *21st Century Science and Technology*, spring 2000, presented in Internet website www.newtonphysics.on.ca.

[32] L. Doyle, H. Deeg, and T. Brown, "Searching for Shadows of Other Planets", *Scientific American*, Sept. 2000, pp. 58-65.

[33] O. Baker, "Planetary potential surrounds most stars", *Science News*, Oct. 9, 1999, vol. 156, No. 15, p. 231.

[34] S. Chandrasekhar, "The Dynamic Instability of Gaseous Masses Approaching the Schwartzschild Limit in General Relativity", *Astrophysical Journal*, vol. 140, No. 2, 1964, pp. 417-433.

[35] Jesse L. Greenstein and Maarten Schmidt, "The Quasi-Stellar Radio Sources 3C 48 and 3C 273", *Astrophysical Journal*, vol. 140, No. 1, July. 1964, pp. 1-34.

[36] Clifford Will, "The Renaissance of General Relativity", in *The New Physics*, Paul Davies Ed, 1989, pps 7-33, Cambridge U. Press, ISBN 0-521-30420-2.

[37] "Plethora of quasars", *Science News*, Jan 23, 1999, p. 57, vol. 155, No. 4.

[38] Ivars Peterson, "A New Gravity: Challenging Einstein's general theory of relativity", *Science News* , Dec. 3, 1994, vol. 146, pps 376-378.

Creation of Life on Earth
[39] Ian Tattersall, "How We Came to Be Human", *Scientific American*, Dec. 2001, pps. 56-63; and *The Monkey in the Mirror: Essays on the Science of What Makes Us Human*, Harcourt, 2002.

[40] "Earliest Evidence of Complex Life", *Science News*, Aug. 28, 1999, vol 156, no 9, p 141.

[41] J. Madeleine Nash, "When life exploded", *Time*, Dec. 4, 1995, vol 146, no 23, pp 66-74.

[42] Richard Monastersky and O. Louis Mazzatenta, "Life Grows Up", *National Geographic*, April 1998, vol 193, no 4, pp 100-115.

[43] Paul F. Hoffman and Daniel P. Schrag, "Snowball earth", *Scientific American*, Jan. 2000, vol 282, no 1, pp 68-75.
[44] R. Monastersky, "Waking Up to the Dawn of Vertebrates", *Science News*, Nov. 6, 1999, vol 156, no 19, p 292.
[45] Kerri Westenberg, "From Fins to Feet", *National Geographic*, May 1999, pp. 115-126.
[46] Richard Monastersky, "Out of the swamps", *Science News*, May 22, 1999, vol 155, no 21, pp 328-330.
[47] Michael J. Benton, *The Reign of the Reptiles*, 1990, Crescent Books, NY, ISBN 0-517-02557-4.
[48] Michael J. Benton, *The Rise of the Mammals*, 1991, Crescent Books, NY, ISBN 0-517-02561-2.
[49] William K. Hartmann and Ron Miller, *The History of Earth*, 1991, Workman Pub, NY, ISBN 0-89480-756-0, p. 176.
[50] *Atlas of Life on Earth*, Barnes and Noble Books, 2001, ISBN 0-7607-1957-8.
[51] Peter Ward, "The greenhouse extinction", *Discover*, Aug, 1, 1998, vol 19, pp 54-55; and, "Mass extinction: the big heat", *The Economist*, Aug. 28, 1999, vol 352.
[52] Christopher P. Sloan and J. Louis Mazzatenta, "Feathers for T. Rex", *National Geographic*, Nov. 1999, vol 146, no 5, pp 98-107.
[53] D. M. McLean, "A terminal Mesozoic 'greenhouse': lessons from the past", *Science*, 1978, vol 201, pp 401-406.
[54] Dewey McLean, "Dinosaur Volcano Greenhouse Extinction", May 2000, Virginia Polytechnic Institute website, www.fbox.vt.edu, search for "Dewey McLean".
[55] Richard Leakey and Roger Lewin, *Origins Reconsidered*, 1992, Doubleday, NY, ISBN 385-46792-3.
[56] Roger Lewin, *The Origin of Modern Humans*, 1993, Scientific American Library, W. H. Freeman, NY, ISSN 1040-3213 (Ch. 1).

History of Gravity and Light (Chapters 5 and 6)
[57] Will Durant, *The Reformation*, 1957, MJF Books, NY, ISBN 1-56731-017-6, pps 855-863.
[58] Will and Ariel Durant, *The Age of Reason Begins*, 1961, MJF Books, NY, ISBN 1-56731-018-4, pps 584-611.
[59] Will and Ariel Durant, *The Age of Louis XIV*, 1963, MJF Books, NY, ISBN 1-56731-019-2, pps 531-547.
[60] Hal Hellman, *Great Feuds in Science*, 1998, John Wiley, NY, ISBN 0-471-16980-3, Ch. 1.
[61] I. B. Cohen, *Arch. Intern. d'Histoire des Sciences*, Oct-Dec. 1958, vol 11.
[62] Isaac Newton, *The Principia*, English transl. by Andrew Motte, 1995, Prometheus Books, ISBN 0-87975-980-0.
[63] Isaac Newton, *Opticks*, Dover Pub, NY, 1952, Library Congress, 52-12165.
[64] David Halliday and Robert Resnick, *Physics*, 1966, John Wiley, NY, Library Congress, 66-11527 (p. 1074).

Books on Relativity
[65] L. D. Landau and E. M. Lifshitz, *The Classical Theory of Fields, vol. 2*, 1973, Butterworth and Heinemann, Oxford, England, (first Russian ed. 1951) ISBN 0-7506-2768-9.
[66] Albert Einstein, *The Meaning of Relativity*, Princeton University Press, 5th ed., 1953, ISBN 0-691-02352-2, (1st ed. 1921), (*See* appendix for 2nd ed., 1945, p. 129).

INDEX

Numbers in brackets [ ] are Bibliography references.

absolute differential calculus, 128
acceleration, 83, 87--96, 126-137
acceleration of gravity, 83-96, 134-137, 146, 231, 235
aether, 106, 109-114, 126, 207
age of earth, 2, , 5, 14 ,44
age of sun, 4, 5
age of stars, 6, 45-50, 62, 171
age of universe
   apparent age, 62, 66-67, 71-72, 165, 198-200, 206
   true age, 5-8, 62, 171-172, 188-189
   age dilemma, 5-6, 63, 171-172
Alfven, Hannes, 42-43,170-175
algae, 15, 19, 21
Alpher, Ralph, 77
Alley, Carroll O, 180-186, [Y10, Y12]
amphibian, 18-19, 21
Andromeda M31 galaxy, 38, 60, 64-65
angular momentum, 40-42
archaea, 14-15
Arp, Halton, 80-81, 161-168, 170, 177-179, 218 [16, 17]

bacteria, 14-15, 21
big bang theory, 2, 5-6, 8-12, 58, 62-63, 66-79, 82, 133, 149-150, 156, 169-179, 186, 195-198, 200-205, 219-220, 224-225, 239
birds, 18, 20-21, 23-25

blackbody radiation, 56-58, 77-79, 172, 196, 213, 220-221, 223-224
   see also, cosmic microwave radiation
black hole, 10-11, 52-58, 73, 75-76, 151, 154-159, 168
   Einstein refutation of, 53-56
Bondi, Hermann, 8, 71, 78-79
Brahe, Tycho, 86
Burbidge, Geoffrey R., 78, 169-170, 177, [20, 27]

calculus, 83-84, 92-94, 131, 146, 208, 231-232
calculus, absolute, 131
Cavendish, Henry, 89-90, 95-96, 182
Cepheid variable stars, 64-65
Chandrasekhar, S., 159, [34]
clock rate, change of,
   special relativity, 123-126, 125, 151, 181, 207
   general relativity, 136-137, 147, 151, 157-159, 181, 186, 209, 231
   Yilmaz cosmology model, 188-191
conservation of energy and matter, 184-185, 202, 238-239
coordinates, 125-130, 138-141, 147, 181, 186, 191, 209, 229, 234, 238
   four dimensional, 209, 229
   proper, 125-128, 191
   rectangular (Cartesian) 96-99, 147, 186, 229, 234
   spherical, 96-99
Copernicus, 84-86

cosmology theories,
  see universe, models of
cosmic background explorer satellite, COBE, 77-79, 196, 203, 220, 224
cosmic microwave radiation, 6, 12, 58, 76-79, 172, 179, 195-196, 203, 220-224
  see also blackbody radiation
  big bang prediction of, 6, 76-79, 172, 179, 220
  Yilmaz model prediction, 12, 79, 179, 187, 195-196, 203, 220-224
  Bell Labs meas, 76-78, 179, 195. 220
  COBE satellite measurement, 77-79, 196, 203, 220, 224
covariance, principle of, 126-127, 129
creation of
  matter, 8, 11-12, 195, 198-202
  earth, 40-43
  life, 14-33
  solar system, 40-43
  stars, 34-58
  sun, 40-43
  universe, 59-82
critical mass density, 224-225
curvature of space, 131-133, 140-142, 183-184

dark matter, *see* matter, dark
death of sun, 43-46
density of universe,
  see universe, density
density of matter, *see* matter, density
Descarte, Rene, 97
de Sitter, Willem, 70
diagonal tensor, 138-139
Dicke, Robert, 77
dinosaur, 22-25
Doppler wavelength shift, 5, 80, 115, 130-137, 160, 166, 193, 196, 203, 216, 220-221, 223, 233, 236-237
dwarf star,
  white, 44-45, 50-51
  black, 45-46, 68, 199
  brown, 35, 68

earth, creation, 40-43
Einstein, Albert, 11-13, 53-56, 70-71, 75-76, 93, 117-119, 128-137, 143, 145, 146-150, 154-155, 178, 180, 182, 207-211, 229, 237-239 [9, 10, 13, 66]
  Einstein rejection of singularities 11, 53-55, 62, 70-73, 75, 76, 149, 154-155, 180, 239,
  see also, Einstein relativity theories
Einstein general relativity, 8-13, 53-56, 62, 70-73, 75-76, 79, 82, 93, 131-159, 166, 174-176, 178-187, 191, 196-199, 206, 208-209, 230-231, 237-239
  See also, Einstein general relativity properties
Einstein general relativity properties
  see also, relativistic effects,
  arbitrariness of, 185, 196
  clock rate, 136-137, 157
  computer studies of, 10, 79, 145, 150
  curvature of space, 131-132, 138, 140, 183-184
  nonphysical predictions, 53-56, 72-76, 154-156
  pseudo-tensor, 148, 237-238
  single-body solution, 180-184
  spatial contraction, 157
  speed of light, 131, 137, 153-157
  tensors, 131, 137-145, 183-184, 226-229
  verification of, 132-133
Einstein gravitational field equation, see gravitational field equation
Einstein photo-electric effect, 118, 210
Einstein relativity theories, 8-13, 53-56, 62, 70-73, 75-76, 79, 82, 93, 117-159, 166, 174-176, 178-187, 191, 197-199, 206-209, 221, 229-232, 236-239
  see also, Einstein general relativity
    Einstein special relativity
Einstein special relativity, 117-130, 176, 207-208, 221, 232, 236-237

Einstein *unified field theory* search, 150, 211
electron, 42, 45, 50-51, 56, 102-103, 126-127, 167, 173, 183, 198, 211, 215-216, 218
electromagnetic wave, 100-104, 112-113, 199, 210
electric field, 102-104
electromagnetic field equation, 112-113, 117, 130, 150, 211
energy-plus-matter conservation, 184-185
energy-to-matter conversion, 12, 195, 199-202
equivalence, acceleration and gravity, 134-137
ether, *see* aether
eukaryote, 14-15
event horizon, 54-55, 155
eye glasses, 105

field equation, gravitational, *see* gravitational field equation
field equation, electromagnetic, *see* electromagnetic field equation
Filkin, David, 72-73, 155-156, [19]
fish, 17-19, 21, 23-24
Fischer, J. R., 171-172
FitzGerald, George, 116
Folsing, 54, 210, [23]
Freedman, Wendy, 65
Friedmann, Alexander, 71
Fresnel, Augustin, 111`-112
fission, nuclear, 43, 53
fusion, nuclear, 4, 35, 40, 43-46, 50-51, 201, 205

galaxies,
  M31 and M33, 38, 60, 64-65
  M51 Whirlpool, 1, 34, 38, 60,
galaxy, Milky Way, 1, 4, 34, 38-39, 59-60, 64, 67-68, 80, 160, 165, 177, 217-218, 224
Galileo, 83-87, 105, 174

Gamow, George, 11, 69-77, 82, 179, 197-198, [14, 15]
Geller, Margaret, 172
geocentric theory, 84
geodesic equations, 131, 191,194, 196
geological period, 33
  Cambrian, 15-17, 22, 49
  Carboniferous, 20
  Cretaceous, 20, 23-25
  Ediacara, 16
  Jurassic, 23
  Permian, 22-24
  Triassic, 22-24
Gold, Thomas, 8, 71, 78-79
Goldsmith, 50, 52, [24]
gravity, acceleration of,
  see, acceleration of gravity
gravity wave, *see*, wave, gravitational
gravitational constant G, 89-96, 144, 152, 182, 224
gravitational field equation, Einstein 9-10, 13, 53, 55, 70, 72, 75, 79, 82, 132-133, 141-145, 154-155, 174, 180-187, 197-198, 209, 237-238
gravitational field equation, Yilmaz 147-148, 183
gravitational potential, 231-236
gravitational redshift,
  see, redshift, gravitational
gravitational theories,
  Einstein, see Einstein gen. relativity
  Newton, *see*, Newton gravity theory
  Yilmaz, *see*, Yilmaz gravity theory
Greenstein, Jesse L., 79-80, 167, 215-219, [35]

Hawking, Stephen, 72-74, 155-156 [19]
helium, 4, 40, 43-46, 50, 129
heliocentric theory, 84-85
Herman, Robert, 77
Hertz, Heinrich, 112-113
Hooke, Robert, 109, 111
Hoyle, Fred, 8, 71, 78-79, 177, [20]
Hubble, Edwin, 1, 4-5, 12, 59-65, 70, 191

Hubble expansion, 1, 4-8, 12, 59-65, 70-72, 78, 80, 165, 172, 191-192, 194-195, 202, 206, 213-214, 224
Hubble constant, 5, 7-8, 61-67, 72, 80, 165, 188-189, 192, 214, 216
Hubble Space Telescope, 61, 64
Huchra, John P., 172
Huygens, Christiaan, 107
human evolution, 25-33
human evolution theories
  language, 3-4, 28-32
  multi-regional, 31
  replacement, 31
humans, related species,
  australopithecus, 25-26
  homo erectus, 26-31
  homo habilis, 26-27
  homo sapiens, archaic, 28
  modern humans, 28-33
  Neanderthal man, 28-32
hydrogen, 4, 7-8, 15, 40, 43-46, 50, 66, 95, 129, 166, 192, 195, 198-199, 202, 208, 213-219, 224-225

ice age, 16, 32
indices in tensors, 138-141, 183-184, 226-229
infinite mass density, *see* singularity

Kepler, Johannes, 83-84, 86, 90-93, 105, 174

language development: 3-4, 28-32
Landau, L. D., 238, [65]
Lemaitre, Georges-Henri, 71
Lerner, Eric J., 42-43, 78, 170-177, [18]
Leavitt, Henrietta, 64
Leibniz, G. Wilhelm, 93-94
Levi-Civita, Tullio, 131, [5]
life on earth, 2-4, 14-33, 178, 205-206
life in universe, 39-40, 47-50
Lifshitz, E. M., 238, [65]
light rays , bending, 132-133

light, speed of,
  measurement of, 120-122
  constancy of, 115-125
  variation with acceleration, 131-132
  variation with gravity, 131-132
light, theory of,
  aether, concept of, 113-117
  corpuscular, 104, 109-110
  electromagnetic, 101-104
  photon,
  wave, 100-104, 109-113
Local group of galaxies, 60-61
Lorentz, Hendrick, 116-118, 123, 129-130, 207 [8]
Lorentz transformation, 116-117, 123, 128-130

Magellanic clouds, 60, 64
magnetic field, 42, 52, 100-104, 112-113, 117, 130, 150, 168, 173, 199, 210-211
magnitude, absolute, 46-50, 58, 64, 217
magnitude, stellar, 46-50, 64, 80, 163-165, 217
mammals, 18, 21-25
Marconi, Guglielmo, 112-113
Marmet, Paul, 2, 7, 166-168, 202, 213-214, 218-219, [30, 31]
Marmet redshift, *See*,
  redshift, Marmet effect
Mars parameters, 69, 75
mass-energy conversion, 126-129
mass, normalized relativistic 152-153, 225, 231, 234
mass density of universe, 7-8, 202, 224-225
  critical mass density, 224-225
  Yilmaz prediction, 224-225
matrix, 138, 228-229
matter, dark, 67-68, 72, 224-225
matter, density of, 224-225
matter and energy conservation, 184-185
matter-to-energy conversion, 128
Maxwell, J. Clerk, 112-113, 115

Maxwell's electromagnetic field eqs., 111-113, 117, 130
measurement for star and galaxy
  distance, 63-66
  velocity, 5, 236-237
Messier, Charles, 60
Mercury, orbit of
  relativistic effect, 133, 181-184, 191
  single-body solution, 181-184
meteorite, 14-15, 22-25, 205
metric tensor, *See* tensor, metric
Michelson-Morley experiment, 115-116, 122
Milky Way, *see* galaxy, Milky Way
momentum, 83, 127-128, 210
momentum, angular, 40-42
multi-body solution
  of Einstein theory, 180-184
myth and cosmology, 172-179

Narlikar, Jayant V., 78-79, 177, [20]
nebula, 1, 7, 52, 59-60, 159, 166-167, 173, 215-219
nebula, gaseous, 7, 159, 166-167, 215-219
neutrino, 51
neutron, 11, 50-52, 55-56, 68-75, 82, 197, 199
neutron star, 11, 50-52, 55, 68-75, 82, 197, 199
Newton, Isaac, [62, 63]
  see, Newton's gravity, optics research
Newton's gravity theory
  82-96, 109, 127-128, 132-133, 143, 152, 181-182, 194-195, 209, 224
Newton's optics research, 105-112

observable universe
  big bang theory, 66-77, 82, 203, 225
  Yilmaz cosmology model, 203-204, 206, 225
Oppenheimer, J. Robert., 52-56, 154, 180, [12]

parallax, 47, 63-64

parsec, 46-47, 216-217
Pauli, W., 45, 51, 55-56, 237, [4]
Pauli exclusion principle, 45, 51, 55-56
Peacock, John A., 35, 149, [6]
Peebles, P. J. E., 9, 74-75, 77, 173, 197, [28]
Penrose, Roger, 72-74, 155-156, [19]
Penzias, Arno, 76-79, 179
photoelectric effect, 118, 210
photosynthesis, 15, 21, 205
photon, 7, 118, 191, 202, 210, 213-214, 221-223
plants, terrestrial, 15, 19-21
plasma, 42-43, 170, 173, 196, 220
proton, 50-51
Proxima Centauri, 38
pseudo-tensor, 148, 237-238
pterosaur, 23-25
pulsar, 52

quantum mechanics, 45, 150, 210-211
quasar, 7, 79-82, 151, 159, 160-168, 177-178, 214-219, 225

radiation, blackbody,
  see blackbody radiation
radio wave, 52, 79-80, 101-102, 112-113
redshift
  velocity (Doppler), 5, 61, 134-136, 215-218, 236-237
  gravitational, 133-137, 146, 158-159, 166-168
  Marmet effect, 7, 166-168, 202, 213-214, 218-219
relativistic effects
  clock rate, 123-124, 136-137, 151, 157-158, 188-190, 230-231
  spatial contraction, 123-125, 137, 151, 157-158, 188-191, 230-231
  speed of light, 137, 151-157, 188-190, 230-231
  synchronization, 123-125
  wavelength, 134-137, 158-159, 231-233

Index 253

relativistic effects, gravitational
  clock rate, 157, 188-191
  Hubble expansion, 191-195
  spatial contraction, 157, 188-191
  speed of light, 153-157, 188-191
  wavelength, 158-159
relativistic effects, velocity
  clock rate, 123-126
  spatial contraction, 123-126
  simultaneity, 123-126
relativity theory, *see*
  Einstein relativity theory
  Yilmaz gravity theory
reptiles, 21-25
  Diapsid, 21-24
  Synapsid, 21-24
Ricci, Gregorio, 131-132, 208
Riemann, Bernhard, 131-132, 208

Sandage, Alan, 65
Schmidt, Maarten, 79-80, 167, 215-219, [35]
Schwartzschild, Karl, 53, 131, 145
Schwartzschild-Einstein solution, 52-53, 131, 143-145, 151-159, 181-185, 191, 230-231
  derivation, 143-145
  event horizon, 54-55, 154-155
  Schwartzschild limit, 53-55, 154-157, 180
  Schwartzschild radius, 154-155
  Schwartzschild singularity, 52-55, 75, 154-155
Silk, Joseph, 52, 67, 73-74, [21, 22]
singularity
  black hole, 10-11, 52-56, 75-76, 154-157, 180, 198
  Big Bang, 9, 11, 72-76, 156, 186, 197-198, 239
  concept, 8-11, 52-56, 72-76, 149, 155-156, 180, 186, 239
simultaneous events, 123-131
Snyder, H., 52-56, 154,180, [12]
solar system, *See*, creation of

sound,
  speed of, 113-114, 119-121
  wave concept, 100, 104
spatial contraction
  from gravity, 53, 157, 230-231
  from velocity, 116-117, 123-125
  Yilmaz cosmology model, 188-191
speed of light, *see* light, speed of
spectral lines, forbidden, 159, 167-168, 215-218
spectral/spectrum, 5, 56, 58-64, 77-79, 101, 106-107, 110, 133, 137, 151, 159-160, 167-168, 172, 196, 215-221
spectrum, blackbody, 56-58
stars
  age of, 6, 45-50, 62, 171
  distance measurement, 63-66
  great distance to, 36-39
  velocity measurement, 5, 236-237
stars, variable
  Cepheid, 64-65
  RR Lyrae, 65
  quasar variation, 80, 160, 167, 218
steady-state universe theory
  see, universe. models of
stress-energy tensor, *see* tensor type
sun
  creation of, 40-43
  life and death of, 43-47
  normalized mass, 152-153
supernova, 50-51, 65, 81, 161
synchronous clocks, 124-126

Tattersall, Ian, 3, 29-32, [39]
telescope, 38, 59-65, 79, 85, 87, 102, 105-106, 160, 170, 178
telescope types,
  field glass (Galileo), 105
  inverted image (Kepler), 105
  reflecting (Newton), 106
  achromatic lens, 105-106
temperature, blackbody, 56-58, 77-79, 196, 220-224
temperature, stellar, 4, 35, 40, 43-44, 50, 159, 215

tensor analysis, 131-132
tensor, basic concept, 137-143, 226-229
tensor, diagonal, 139
tensor forms of, 139-140
 (contravariant, covariant, mixed)
tensor, types of
 Einstein, 138, 140, 183-184
 energy-momentum, 138, 141, 147-148
 metric, 138, 139
 pseudo-tensor, 148, 237-238
 Riemann, 183-184
 Ricci, 138, 140, 183-184
 stress-energy for gravity, 147-148
 stress-energy for matter, 147-148
thermodynamics, laws of, 200-203
Tulley, Brent, 171-172

universe age dilemma, 5-6, 63, 171-172
universe, distances in, 63-66
universe, mass density of, *see* mass density
universe, models of,
 big bang, *see* big bang theory
 Marmet theory, 7, 213-214
 steady-state universe, 8, 78
 Yilmaz cosmology model, 187-196, 220-225
universe radius, big bang
 observable, 66-70, 225
universe radius, Yilmaz
 apparent, 189-191
 observable, 203-204
universe, structure of, 171-172
 (filament and ribbon),
unified field theory, 150, 211

vector, 91, 96-99
vertebrates, 17

wave, electromagnetic, 101-104, 109-118
wave, gravitational, 199-200
wave, light, 101-104, 108-118
wave, sound, 100, 113-115, 124
wave, water, 100
Wilson, Robert, 76-79, 179
Wolpoff, Milford, 31

Yilmaz, Huseyin, 10-13, 136, 146-150, 181, 185, 187, 209-210
Yilmaz cosmology model. 8, 11-12, 79, 179, 187-206, 220-225, 230-239
 clock rate, 188-191
 creation of matter, 195, 199-203
 cosmic microwave radiation, 179, 195-196, 220-224
 Hubble expansion, 191-195
 matter in universe, 203-204, 224-225
 spatial contraction, 188-191
 speed of light, 188-190
 uniqueness, 185, 196
 universe density, 203-204, 224-225,
Yilmaz gravity theory, 8-11, 13, 55-56, 136, 146-150, 151-159, 166, 181, 183, 185, 186, 209-211
 derivation, 146-148, 231-237
 clock rate, 157
 gravitational field equation, 147-148, 183, 185, 237-238
 spatial contraction, 157
 speed of light, 153-154
 stress-energy tensors,
  for matter, 147-148, 185, 238-239
  for gravity field, 147-148, 183
 wavelength, 158-159
 quantum mechanics, 150, 211
 uniqueness of, 185, 196
 time-varying solution, 148
Young, Thomas, 107-109